WATER PRODUCTIVITY IN RAINFED AGRICULTURE

Redrawing the rainbow of water to achieve food security in rainfed smallholder systems

WATER PRODUCTIVITY IN RAINFED AGRICULTURE

Redrawing the rainbow of water to achieve food security in rainfed smallholder systems

DISSERTATION
submitted in fulfilment of the requirements of
the Board for Doctorates of Delft University of Technology
and of the Academic Board of the UNESCO-IHE Institute for Water Education
for the Degree of DOCTOR
to be defended in public
on Wednesday, 23 June 2010 at 12.30 hours
in Delft, The Netherlands

by

Hodson MAKURIRA
born in Bindura, Zimbabwe
Master of Science in Water and Environmental Resources Management
UNESCO-IHE, Delft, The Netherlands

CRC Press
Taylor & Francis Group
Boca Raton London New York

CRC Press is an imprint of the
Taylor & Francis Group, an **informa** business
A BALKEMA BOOK

This dissertation has been approved by the supervisors:
Prof. dr. ir. H.H.G. Savenije
Prof. dr. S. Uhlenbrook

Committee members:

Chairman	Rector Magnificus TU Delft
Prof. dr. A. Szöllösi-Nagy	Vice-chairman, Rector UNESCO-IHE
Prof. dr. ir. H.H.G. Savenije	TU Delft/ UNESCO-IHE, The Netherlands
Prof. dr. S. Uhlenbrook	UNESCO-IHE/ TU Delft, The Netherlands
Prof. dr. D. Mazvimavi	University of the Western Cape, South Africa
Prof. dr. ir. P.van der Zaag	UNESCO-IHE, The Netherlands
Prof. dr. W. Bastiaanssen	TU Delft, The Netherlands
Prof. Bob Su	University of Twente, The Netherlands

Reserve member	
Prof. dr. ir. T.N. Olsthoorn	TU Delft, The Netherlands

Published by:
CRC Press/Balkema
Schipholweg 107C, 2316 XC Leiden, The Netherlands

ISBN 13: 978-0-415-60120-7 (pbk)

ABSTRACT

The challenge of food insecurity is severe in sub-Saharan Africa (SSA) where, generally, low and highly variable rainfall coupled with high evaporation rates and rapidly degrading soils combine to produce declining grain yields. Soil moisture and nutrient balances are essential ingredients for good crop yields. Water scarcity is perceived to be the most limiting factor to crop productivity (CP) in SSA. With a projected population of 1.1 billion in 2010 which is growing at a rate of 2.2 % a^{-1}, SSA will continue to face the challenges of hunger and poverty for the foreseeable future unless if interventions are made to control the situation.

In SSA the seasonal grain yields are low with average maize grain yields, for instance, hardly exceeding 1 t ha^{-1} which is very low when compared with potential levels of 6 t ha^{-1}. The solution to water shortages for agriculture has traditionally been advanced as irrigation. However, irrigation has only benefited a few owing to the high investment costs required to set up irrigation schemes thus leaving the majority (80-90%) of the population in SSA relying solely on rainfed agriculture for their livelihoods.

Rainfall is generally unreliable in SSA with seasonal rainfall ranging between 300-1200 mm a^{-1}. Within seasons, rainfall variability is very high. So is potential evaporation which easily exceeds 1000 mm a^{-1} in many places. This implies that every little available raindrop should be converted to productive use to improve CP.

The challenge of water scarcity as a result of insufficient seasonal rainfall and dry spell occurrences during seasons is compounded by inefficient agricultural practices by smallholder farmers where insignificant soil conservation efforts are applied. When these dry spells occur at critical growth stages, significant yield reductions occur even where the total seasonal rainfall may be considered good.

The farmers also lack resources to purchase the most appropriate farming inputs which also help the crops to cope with water and nutrient deficits. Interestingly, a lot of research has gone into coping mechanisms against dry spells, soil and water conservation and farm management. However, on the ground the situation has not improved and, if anything, yields continue to decline.

There appears to be a problem with research, so far, in offering sustainable solutions to the persistent low crop water productivity in smallholder rainfed farming systems. This research tackles this problem. The hypothesis of this research is that many of the past research efforts have taken a fragmented approach to deal with the challenges highlighted here. Fragmented approaches are difficult to implement. A holistic approach to assist traditional farming systems should include hydrological aspects, agronomy, soil science, catchment conservation and socio-economic aspects for better success.

In this research the Makanya catchment in northern Tanzania has been taken as a pilot study site. The area receives gross rainfall of below 400mm/season which is clearly insufficient to meet water requirements of the preferred crops such as maize. System innovations (SIs) have been introduced and tested as alternative farming practices. These include a combination of conservation agriculture, diverting runoff onto field plots and enhancement of in-field soil moisture through trenching and soil bunding (*fanya juus*) within cultivated fields plots. These techniques have been selected for study because they do not require large capital investments and, hence, are affordable to many farmers yet they have the potential to significantly improve yields.

Comprehensive on-site observations of rainfall, soil evaporation, runoff contribution, seasonal grain yields of the maize crop and general crop performance have been conducted using a participatory approach with local smallholder farmers.

Indirect methods have been applied to confirm these observations and also, to model the performance of the studied system. Electrical resistivity tomography (ERT) has been applied as a geophysical technique to confirm the observations from the Time Domain Reflectory (TDR) methods of monitoring soil moisture. The HYDRUS2D model has also been applied to simulate two-dimensional sub-surface lateral moisture flows in relation to applied water on the field site with conservation agriculture. A spreadsheet based water balance model has been applied to better understand the water partitioning processes under different scenarios and to quantify crop productivity.

Results from the research show that rainfall is generally low and ranged between 150 - 300 mm/season during the research period. Rainfall variability is high between seasons and between the studied sites. The maize grain yields obtained are also low and range from an averaged minimum of 0.23 t ha^{-1} under the current agricultural practices to a maximum of 2.82 t ha^{-1} when SIs were applied at all sites. Average yield values show that, in addition to the rainfall received, a combination of diversion of runoff onto field plots, ripping along planting lines, promotion of soil moisture storage by use of *fanya juus* and the application of manure produced the best yield results of up to 4.8 t ha^{-1} in a good season at one site. Water partitioning analysis has confirmed that the SIs applied resulted in an increase in transpiration values of about 49%. An explanation on the reasons for this increased efficiency has been obtained by reference to both direct and indirect analytical techniques.

The research has successfully applied different analytical techniques to better understand soil and water interactions at field scale. It has been successfully demonstrated that there is indeed scope to increase crop water productivity provided the local farmers adopt more efficient cultivation techniques. Significant yield increases occur as a result of diverting more water and these further improve when other SIs such as ripping, application of manure and cover cropping, are introduced. This confirms that no single solution exists to solve the problem of low yields which are obtained in smallholder farming systems. The SIs that have been tested offer improved food security through increased grain yields and also facilitate alternative cropping within the field as a result of heterogeneous conditions which are created within the field as a result of alteration of the field water balance.

However, even with these promising results, the research has shown that there is room to further improve the efficiency of crop water use through improvement in research approaches and exploration of better techniques.

ACKNOWLEDGEMENTS

When a former academic supervisor steps forward to offer to supervise you "again" at a higher level, it gives one a sense of confidence and pride that one didn't do such a bad job after all during the first contact. I would like to express my since appreciation to Professor Huub Savenije for giving me that sense of confidence after guiding me through my MSc research so many years ago. I will always admire your total commitment to see the best out of your students.

To Johan Rockström for setting the tone for this research and all the support during those early stages. I hope you will be proud of this final product.

To Stefan Uhlenbrook, you came at the right time when the morale was low and the ship seemed to be losing direction. I will always treasure your valuable efforts to keep the product scientifically sound.

To Aiden Senzanje, last but not least, of the supervisory crew. You gave me constant support and those encouraging remarks, often off-air, kept me going.

This work is a product of part of a complex multi-disciplinary research. My sincere thanks to all those who formed part of the SSI research team at one point or another. These include Marloes Mul, Victor Kongo, Elin Enfors, Job Rotich, Kenneth Masuki, Claudious Chikozho, Jenniffer Kinoti, Jeltsje Kemerink, Charles Hans Komakech, Jayashree Pachpute, Siza Tumbo and Line Gordon. I also benefitted a lot from interactions with staff and fellow PhD'ers at UNESCO-IHE and TU Delft.

I had contact with many MSc researchers from University of Zimbabwe, Sokoine University, TU-Delft, UNESCO-IHE, Stockholm University, Stuttgart University and

University of Freiburg. Thier direct and indirect contribution to my work is highly valued.

To Eliza, Walter, Iddi, Wilson (may his soul rest in peace) for providing me with the research sites and participating in the research. You all had your unique contributions which helped to shape this research. Equally important was the contribution from Maliki, Msangi and Kapombe who served as the research assistants.

The initial days of my work were challenging and I wish to acknowledge the tremendous support received from Professor Pieter van der Zaag and Professor Simbi. I will never forget those big fights with the administration of which, in my view, there was no outright winner.

To Joseph Mwalley who came in with the idea of the *fanya juus* and helped to set them up. I hope the *Lexus* is still on the road.

My profound appreciation to the Soil Moisture Group of the Sokoine University of Agriculture for hosting this research. Equally, I wish to express my gratitude to the University of KwaZulu Natal for affording me the opportunity to acquire more depth. Special mention goes to Simon Lorentz (instrumentation) and Vincent Chaploit (data analysis).

To colleagues who have walked the path before me and gave me tremendous support. These include Marieke de Groen, Innocent Nhapi, Themba Gumbo, and Lawrence Nyagwambo

I also owe this product to the entire Civil Engineering staff compliment who covered up for me during the times when I was away on "academic duty". Zvikomborero Hoko, your support is highly acknowledged.

My research collaborated with the Challenge Programme in the Mzingwane Catchment. I would like to encourage my colleagues who are still walking the walk (David, Alex and Collin) to keep on keeping on.

This research would not have taken off without the financial support which came from SIDA, DGIS, IWMI and UNESCO-IHE.

Marloes, I owe you so much during our long and memorable journey in the land of the waPare. *Ubarikiwe sana.* Elin, I appreciate the thorough proof reading and standing next to me up to the last day Who else could do it other than my sister? To Jeltsje and Miriam thanks for all especially the last minute efforts to get this document in shape. I witnessed the efficiency of lady power during the last days of this work.

I would like to acknowledge the morale and social support provided by all colleagues at various circles back home. The list is endless but I will always treasure every moment of support.

Lastly, a big thank you to the close family, my mother, Lulu, Keith and Mavie for bearing with me during the time that I could not be with you. I promise to spend more time with you guys after this.

PREFACE

The year was 2003 and I had just joined the University of Zimbabwe as a lecturer in the Civil Engineering Department. Huub Savenije was visiting the University of Zimbabwe on his annual pilgrimage to lecture on Principles of Hydrology to the current Masters' class. During discussions he mentioned something about a research proposal which was almost getting approval and how it would be good if I became one of the members of the promising "Dream Team" to participate in a multi-disciplinary research. I did not take those remarks seriously until, some weeks later, Johan Rockström invited me to his office for a "chat" where he confirmed that research funding had been obtained for an integrated research involving 8 PhD and 2 Post doctoral studies in two basins in Tanzania and South Africa. After a bit of hesitancy and extensive consultations with friends and family, I finally embarked on this long journey of PhD research within the SSI programme.

The rest, as they say, is history.

This research has opened my eyes as far as smallholder rainfed agriculture is concerned. So often we blame "bad" seasons as the reason for endless drought-induced famines especially in semi-arid regions like where I come from. But what is a bad season? There is more to describing a "good or bad" season than the cumulative seasonal rainfall totals that we often use as a measure of the quality of a season. Farm management practices, the choice of crop and seed variety, the distribution of rainfall in a season and, as my research showed, how a farmer tampers with the water balance at field scale through runoff management can help to concentrate moisture within the root zone. Water is a major limiting factor to crop productivity. Of course, farm management practices, including timing and soil and nutrient management are equally important.

The research work took me far and wide and I made friends and family across many regions. I fell in love with Tanzania and the people. I also got to appreciate, better, diverse cultures and diverse personalities. I benefited a lot from being part of this multi-disciplinary team. Mama Ntilie proved to be a mother away from home and her family was my family notwithstanding the fact that she could not tell the difference between me and a beautiful Dutch blonde girl. I felt so much at home at the Pangani Basin Office, thanks to Mzee Macha and Mzee Basso. Technically the engineer at Same, Fundi Mbonea, was always available to fabricate the" rain-makers", (tipping buckets),the lysimeter and any other gadget I needed and still he had time for a *baridi sana* once in a while.

The fieldwork was more than an academic exercise. What with all those diverse discussions in the Landrover up and down the mountains? It was so much fun with drivers Hamadi, Mtweve and Abdallah.

Despite all these fond memories of the good times in Tanzania, I realise that my research area has been quite challenging as it is, in all honesty, not my forte. So often people asked me what my research is all about and what field it falls into. I never had a straight answer and even now I don't think I have. This thesis is neither pure hydrology nor pure agriculture. To me, this makes this product unique as it links hydrology and agriculture (agronomy). More research needs to be undertaken in this area and I feel I have done my part.

I consider myself lucky in having participated in a funded research. I appreciate that not many would fall into this fortunate position, but I will always encourage them to soldier on. A few certainties, however, came out of this experience. First, there is great potential to obtain better yields under existing challenging environmental and climatological scenarios. Second, I have convinced myself that I have acquired some agricultural expertise which equips me to become a successful farmer one day.

Lastly, I have all the confidence that this work will make a big change if these simple and affordable techniques that I have explored in this research are pursued by someone somewhere at some level.

LIST OF SYMBOLS

Symbol	Parameter description	value/ unit	Dimension
A_d	Cross sectional area of lysimeter drum		L^2
b	Reduction scale		L
D	Interception threshold	2 - 5 mm d^{-1}	LT^{-1}
D_r	Root depth		L
E	Evaporation		LT^{-1}
E_I	Evaporation from interception		LT^{-1}
E_f	Nash and Sutcliffe coefficient of efficiency	$-\infty$ - 1	-
E_L	Evaporation from lysimeter		LT^{-1}
E_{ref}	Reference evaporation		LT^{-1}
E_s	Evaporation from the soil		LT^{-1}
E_T	Evaporation from transpiration		LT^{-1}
F	Infiltration		LT^{-1}
f_{ms}	Moisture stress reduction factor (for soil evaporation)		-
f_{mt}	Moisture stress factor (for transpiration)		-
I_{LA}	Leaf area index		$L^2 L^{-2}$
k	Moisture stress gradient		L^{-1}
k_c	Crop factor	0.15 – 1.15	-
k_p	Pan coefficient	0.6 – 0.8	-
k_R	Time allowed for soil moisture to exceed field capacity		T
K_s	Soil evaporation coefficient		-
P	Precipitation		LT^{-1}
p	Fraction of no moisture stress	0.6	-
Q_{est}	Estimated discharge		$L^3 T^{-1}$
Q_g	Groundwater flow		$L^3 T^{-1}$

Q_{max}	Peak discharge		$L^3 T^{-1}$
R	Deep percolation		LT^{-1}
S_{fc}	Soil moisture at field capacity		L
S_{max}	Maximum soil moisture in root zone		L
S_u	Soil moisture storage in root zone		L
S_{wp}	Soil moisture at wilting point		L
T_p	Potential transpiration		LT^{-1}
$T_{p,adj}$	Adjusted potential transpiration		LT^{-1}
ΔV_L	Volume of water added into lysimeter other than rainfall		L^3
ΔW_E	Manual weight added to lysimeter		M
ρ	Density of water		ML^{-3}
Δt	Change in time		T
ΔW	Change in weight observed in lysimeter		M
ΔV	Volume of water added into lysimeter		L^3
$\frac{dS_u}{dt}$	Rate of change of water storage in root zone		LT^{-1}
$\frac{dSs}{dt}$	Rate of change of surface water storage		LT^{-1}
$\frac{dS_g}{dt}$	Change of groundwater storage		LT^{-1}

List of Acronyms

AWC	Available Water Content
CWP	Crop Water Productivity
ERT	Electrical Resistivity Tomography
FAO	Food and Agriculture Organisation
MDGs	Millennium Development Goals
NGO	Non-Governmental Organisation
PBWO	Pangani Basin Water Office
RMSE	Root Mean Square Error
SI	System Innovation
SSA	Sub-Saharan Africa
SSI	Smallholder Systems Innovations in Integrated Watershed Management project
SUA	Sokoine University of Agriculture
TDR	Time Domain Reflectometry
UNFPA	United Nations Population Fund
URT	United Republic of Tanzania
WMO	World Meteorological Organisation
WP	Water Productivity
WSI	Water System Innovation

List of Figures

Figure 1.1 The interaction and integration off SSI projects (SSI, 2002). 6
Figure 2.1 The rainbow of water as conceptualised by Savenije (2000). 10
Figure 2.2 Partitioning of rainwater at smallholder farm scale (Rockström, 2001).. 11
Figure 2.3 Maize productivity levels in selected sub-continents in the past decades (adapted from FAOSTAT, 2005). 13
Figure 3.1 The Makanya Catchment. 24
Figure 3.2 Rainfall trends at nearby Same station for the period 1934-2007 (source: Mul, 2009)... 26
Figure 4.1 A schematic illustration of options for improving agricultural water management in dryland cropping systems (adapted from Nyagumbo et al., 2009). 36
Figure 4.2 The *fanya juu* cultivation technique showing zones of increased infiltration potential. 38
Figure 4.3 Typical experimental set-up showing the location of the treatment plots in relation to the diverted water... 41
Figure 4.4 *Fanya juu* trenches and rip lines across a cultivated strip showing (a) water stored in the trenches after a rainfall event, and (b) much wetter rip lines after a rainfall event at the onset of a cropping season. 42
Figure 4.5 The lysimeter set up at Site 3 to monitor soil evaporation.................. 45
Figure 4.6 The set-up of the *fanya juu* technique with TDR soil moisture monitoring tubes. 47
Figure 4.7 Cumulative seasonal rainfall observed at each research site between January 2005 and April 2008.. 48
Figure 4.8 Daily rainfall received at the research sites during the long rainfall seasons (*Masika*) 2005 -2007.. 49
Figure 4.9 Daily rainfall distribution during the short seasons *(Vuli)* between 2005-2007... 50
Figure 4.10 Typical runoff hydrographs at Site 4 at selected days in the long rainfall season (2006)... 51

Figure 4.11 Comparison of cumulative rainfall (P) received and cumulative runoff diversion (Q) onto the field plots at (a) Site 3 and (b) Site 4 during the periods 2006-2008 52
Figure 4.12 Soil moisture variations at the different locations of the tubes during different seasons. 54
Figure 4.13 Soil moisture variations at different tube locations at Site 3 during all seasons 55
Figure 4.14 Soil moisture variations at different tube locations at Site 4 during all seasons 56
Figure 4.15 Comparison of daily soil evaporation rates measured manually and by use of an automatic strain gauge in 2007. 57
Figure 4.16 Leaf area index for the Kito maize variety as measured in Makanya catchment in 2006 58
Figure 5.1 Flow chart for determining evaporation and transpiration. 66
Figure 5.2 Model results at different sites compared with observed soil moisture values for Site 3 and Site 4. The graph at Site 4(b) presents an improved simulation than the graph at Site 4(a) 72
Figure 5.3 Comparison of observed and modelled moisture in Tube A (control) and Tube D. 74
Figure 5.4 Comparison of seasonal transpiration values for different interception thresholds. 76
Figure 5.5 Pattern of soil moisture variation at Sites 1 and 2 with different field layouts. 81
Figure 5.6 Soil moisture variations with time and location as computed by HYDRUS2D and compared with the output from the hydrological spreadsheet model (SModel) 83
Figure 5.7 Impact of *fanya juus* and diversions at different observation points across the cultivated strip at Site 3. 84
Figure 5.8 Box plots of soil moisture variation between *fanya juu* constructions in all wet seasons during 2006 - 2008 at (a) Site 3 and (b) Site 4. ... 90
Figure 5.9 Box plots of soil moisture variations between *fanya juu* constructions in all dry seasons at (a) Site 3 and (b) Site 4 during the period 2006-2008 91
Figure 5.10 Box plots of soil moisture variation in all wet seasons (2006-2008) at (a) Site 3 and (b) Site 4 for centre tubes 92
Figure 5.11 Box plots of soil moisture variation in dry seasons at (a) Site 3 and (b) Site 4 for centre tubes. 92
Figure 5.12 Absolute resistivities [Ω.m] at Site 3 on 11 April 2006 with dotted lines indicating the location of the *fanya juu* constructions 93
Figure 5.13 Absolute resistivities [Ω.m] at Site 4 on (a) 8 April and (b) 11 April 2006 at Site 4 with dotted lines indicating the location of the *fanya juu* constructions 94
Figure 6.1 Average seasonal yields at each site over four seasons (2006-2007) 100

Figure 6.2 Average grain yields from four seasons per treatment for the four study sites over four seasons (2006-2007). 101
Figure 6.3 Distribution of grain yield per rip line as measured from upslope of terrace at (a) Site 3 and (b) Site 4 in the *Vuli* 2008 season. 102
Figure 6.4 Variation of total water productivity with grain yield at all sites and seasons (2006 - 2008). 104
Figure 6.5 Response of yield per treatment to water availability as observed in all seasons (2006 – 2008). 106
Figure 6.6 Cassava harvest in dry season at Site 3. 107
Figure 7.1 The link between investments, yields and derived benefits. 118
Figure 7.2 Overview of the synthesis of the research. 120

TABLE OF CONTENTS

Abstract v

Acknowledgements ix

Preface xiii

List of Symbols xv

List of Acronyms xvii

List of Figures xix

Table of Contents xxiii

Chapter 1 1

Introduction 1

1.1 Background 1

1.2 The SSI Programme 5

1.3 This research 6

Chapter 2 9

Rainfed agriculture in sub-Saharan Africa 9

2.1 Water and food security 9

2.2 Water productivity 13

2.3 Water for food in sub-Saharan Africa 16

2.4 Scope for improved crop productivity 17

2.5 Research and information gaps 18

2.6 Objectives and research questions 20

2.6.1 Main objectives 20

2.6.2 Key research questions 21

Chapter 3 23

The study area 23

3.1 Physiography 23

3.2 Rainfall 25

3.3 Demography 26

3.4 Soils ---------- 27
3.5 Current land use practices ---------- 27
3.6 Water for food security and coping strategies ---------- 28
3.7 Yields obtained ---------- 30
3.8 Traditional farming preferences ---------- 31
3.9 Discussion and conclusions ---------- 32

Chapter 4 **33**

Research sites and observation techniques **33**

4.1 Site selection ---------- 33
4.2 Background to the development of tested techniques ---------- 34
4.2.1 Conservation tillage 36
4.2.2 Seed selection 39
4.2.3 Cover cropping and manure 39
4.2.4 Rainwater harvesting 39
4.3 Typical experimental setting ---------- 40
4.4 Conceptual framework ---------- 42
4.5 Parameters measured ---------- 43
4.5.1 Rainfall 43
4.5.2 Net in-field runoff contribution *(Q_s)* 44
4.5.3 Soil evaporation *(E_s)* 44
4.5.4 Soil moisture measurements 46
4.5.5 Biomass and leaf area measurements 47
4.5.6 Yield observations 47
4.6 Results ---------- 48
4.6.1 Rainfall 48
4.6.2 Runoff generation 51
4.6.3 Net runoff contribution 51
4.6.4 Soil moisture observations 52
4.6.5 Soil evaporation 56
4.6.6 Biomass measurements 57
4.6.7 Obtained yields 58
4.7 Discussion and conclusions ---------- 59

Chapter 5 **63**

Water partitioning analysis using modelling techniques **63**

5.1 Introduction ---------- 63
5.2 Water balance modelling ---------- 64
5.2.1 Results 71
5.2.2 Water balances 76
5.2.3 Analysis and discussion of results 77
5.2.4 Conclusions 78
5.3 Application of the HYDRUS2D model to interpret sub-surface flow dynamics79
5.3.1 Background to the HYDRUS2D model 79

5.3.2 Model setup and inputs 79
5.3.3 Results 80
5.3.4 Discussion and conclusions 85
5.4 Application of geophysical methods and repeated soil moisture measurements to interpret sub-surface flow dynamics -------- 86
5.4.1 Introduction 86
5.4.2 Direct soil moisture monitoring 87
5.4.3 Indirect soil moisture monitoring (ERT) 88
5.4.4 Data processing 89
5.4.5 Results 89
5.4.6 Discussion 94
5.4.7 Conclusions 95

Chapter 6 **97**

Productivity analysis **97**

6.1 Introduction -------- 97
6.2 Data collection -------- 98
6.3 Data analysis -------- 98
6.3.1 Grain yield 99
6.3.2 Water productivity 102
6.4 Discussion of results -------- 104
6.4.1 Water availability 104
6.4.2 Yields and water productivity 105
6.4.3 Additional benefits of "improved" techniques 106
6.5 Conclusions -------- 107

Chapter 7 **111**

Synthesis of the research **111**

Chapter 8 **123**

Conclusions **123**

References **129**

Samenvatting **139**

About the author **143**

CHAPTER 1

INTRODUCTION

1.1 BACKGROUND

Semi arid zones are generally characterised by rainfall amounts not exceeding 500 mm a^{-1} or whose potential annual evaporation easily exceeds the annual precipitation. Communities in the arid and semi-arid zones of the world are faced with food deficits almost on an annual basis due to crop failure. This has seen millions of people surviving on food relief measures to avert starvation disasters. Food relief measures create a huge burden on governments and relief organisations. In arid and semi-arid regions, crop failure is closely related to a strong imbalance between effective rainfall and high potential evaporation (Hatibu et al., 2006). The resultant sub-optimal yields obtained fall way below the annual food requirements for the ever-increasing world population. The situation has been exacerbated by the general environmental degradation where soil and nutrient loss continue without strict control.

The population of sub-Saharan Africa (SSA) is growing at a rate of about 2.2% a^{-1} and is expected to reach 1.1 billion in 2010 (UNFPA, 2008). Cereal imports in the same region are projected to increase from 9 million t a^{-1} in 1990 to about 35 million t a^{-1} by 2025 (IWMI, 2003). Given that water needs are directly proportional to population and economic growth (IWMI, 2003), the whole of Sub-Saharan Africa is projected to be either economically and/or physically water scarce by 2025 (Falkenmark, 1995). The future of food security in the region looks gloom unless if interventions are made to address the water scarcity situation (Rockström et al., 2009). Food production was observed to stagnate in sub-Saharan Africa in 1970 with grain per capita showing a steady downward trend and by 1984, 24 sub-Saharan African countries recorded abnormal shortages of food with drought blamed for subsequent famine conditions (Pacey and Cullis, 1986). In SSA, these low production levels are still dominating today (FAO, 2005).

Water scarcity poses a threat to food self-sufficiency particularly looking into the future (de Fraiture et al., 2007; Hoff et al., 2009; Kijne et al., 2009). The scarcity of water for food production is a big challenge for water management (Savenije, 1998). 90% of sub-Saharan Africa's population depends on rainfed agriculture for food production (Rockström, 2003b). If food is not produced in sufficient quantities locally, respective governments are forced to import food as a measure to avoid famine disasters. Food imports are common in sub-Saharan Africa and, in a way, indicate a lack of long term strategies to develop mechanisms to cope with water scarcity and climate variability at local scales. Dry spells of between 10 and 21 days occur more frequently during growing seasons than meteorological[1] droughts (Barron et al., 2003). Practically, policy makers and water managers concentrate more on clean water supply projects, which take up a very little percentage of basic human water requirements and on large scale irrigation which accounts for 70-80% of the world's developed freshwater resources (Savenije, 1998). But, whereas dry spell occurrences are the main reason for crop failure, relatively little attention is dedicated to dry spell mitigation.

Given the above scenario, around 90% of the population in sub-Saharan Africa is faced with a high likelihood of annual food deficits unless there is sufficient water to see crops through to maturity. In Malawi 90% of the population rely on subsistence agriculture for their livelihoods while similar patterns are observed throughout

[1] Meteorological droughts were estimated to occur (on average) once in 10 years in this region (Rockström, 2003)

Eastern and Southern Africa (e.g. Botswana (90%), Kenya (76%) and Zimbabwe (70-80%) (Rockström, 2000). The annual rainfall varies between 200 - 1000 mm a^{-1} in the arid to *semi-arid* zones with a mean of 400-600 mm a^{-1} (Ngigi, 2003). This rainfall range is just sufficient for seasonal crop water requirement for common staple food crops and would see crops to maturity e.g. maize (500-800 mm/season), sorghum (450-650 mm/season), beans (300-500 mm/season) (Critchley and Siegert, 1991). However, when evaporation and seasonal variability are taken into account, this average rainfall alone is insufficient to meet many seasonal crop water requirements. This therefore strongly suggests that most of Southern Africa and SSA in general, is vulnerable to food shortages that are caused by agricultural water scarcity.

A number of options exist which can work to reduce the impact of water scarcity on agricultural production. Such options include often costly measures such as relocation, use of treated seeds, large scale irrigation schemes, virtual water trade and the application of fertilisers and pesticides. The viability of these options is highly questionable as social preferences and, often, the capital investments required discourage and, in many cases, prohibit the common rural farmer from considering such options.

In order to achieve self-sufficiency in food production it is obvious that the little amount of rainfall received in rainfed systems should be channelled, as much as possible, towards crop water production. Rainwater harvesting is one viable option to increase the productivity of water and to cushion crops against the impacts of droughts and dry spells (Mupangwa et al., 2006; Mwenge Kahinda et al., 2007). Simple rainwater harvesting technologies are being applied across the sub-continent with varying degrees of success. There is, however, more scope of improvement in rainwater harvesting systems (Hatibu and Mahoo, 1999; Ngigi, 2003; Temesgen et al., 2007). Better and more efficient innovations are needed to increase crop output across the greater part of SSA.

From a scientific research perspective, a huge gap exists in quantifying the incremental benefits which may be derived through improved agricultural systems. This gap creates a weak link between research, outreach on the promotion of promising strategies and the subsequent adoption of such efficient technologies by the intended beneficiaries. Pacey and Cullis (1986) observed that information about existing traditions in runoff farming is inadequate throughout the sub-Saharan region and blames agricultural scientists of contributing next to nothing towards the documentation of the existence of such innovations and their successes. To date, no significant ground has apparently been covered to close this gap. Reij et al. (2001) also acknowledge the lack of information on rainwater harvesting in sub-Saharan Africa and state that the information that is available is extremely limited, fragmented and has not been collected and analysed systematically.

This research is part of a broad and integrated research aimed at uplifting livelihoods in rainfed agricultural systems. The research seeks to establish coping strategies against dry spell occurrences. The study has revealed that a variety of agricultural techniques are being applied at farm level to cushion yields against the impact of dry spells.

It is believed that soil and water conservation technologies, especially rainwater harvesting, will go a long way in alleviating poverty in the region through supplementing the little available rainfall. Incremental benefits are more noticeable at smallholder scale where effective rainfall is generally low, i.e. in the 200 - 600 mm a^{-1} range (Pacey and Cullis, 1986). Such low rainfall areas, which are also often associated with infertile soils, also coincide with settlement patterns of most of the more vulnerable rural population in the sub-Saharan region.

Any efforts towards improved rainwater harvesting technologies are, however, ineffective if there is no careful balance between rainfall, when and where it occurs, and evaporation (Pacey and Cullis, 1986). Effective rainfall, i.e. the part of rainwater which eventually ends up as infiltration and recharges the root zone, is a significant portion of water within the hydrological cycle which plays a crucial part in crop production. The fact that rainwater harvesting techniques are more successful in areas of high water scarcity does not take away the importance of adoption of such strategies in areas of higher rainfall within the sub-Saharan region. Such techniques cushion against the impact of drought and dry spell occurrences even in average rainfall seasons.

Besides rainwater harvesting, other options exist which can also be explored for increased crop productivity. Conservation agriculture, for instance, is an alternative cultivation technique which is different from the traditional hand-hoe technique. Instead, in conservation agriculture, the objective is to minimise the disturbance of the soil structure during cultivation and maximise the water holding capacity. Overturning of the soil during cultivation increases direct evaporation from the soil (Rockström et al., 2001).

1.2 THE SSI PROGRAMME

The Smallholder System Innovations (SSI) Programme (2004 - 2008) has taken up the challenge of balancing water for food and nature. The programme embarked on applied research in semi-arid SSA with the objective of generating knowledge that can be used to uplift livelihoods of vulnerable communities. The research programme took a multi-disciplinary approach involving cross-cutting research on water and environmental processes and management within the context of adaptation and adoption of system innovations focusing on the spatial scale of a catchment.

The main objectives of the SSI programme were:

- to analyse the hydrological, environmental and socio-economic consequences of upscaling water system innovations in smallholder, predominantly rainfed agriculture at catchment scale; and

- to develop methodologies and decision support tools for improved rainwater management and equitable sharing of water between upstream and downstream users and uses in nature and society.

The SSI Programme acknowledged that the solution to the challenges highlighted above do not only require a multi-disciplinary approach, but also should acknowledge the need for active participation by the targeted beneficiaries of the research.

Figure 1.1 shows the interactions between the different projects within the SSI programme in order to achieve the general research objective.

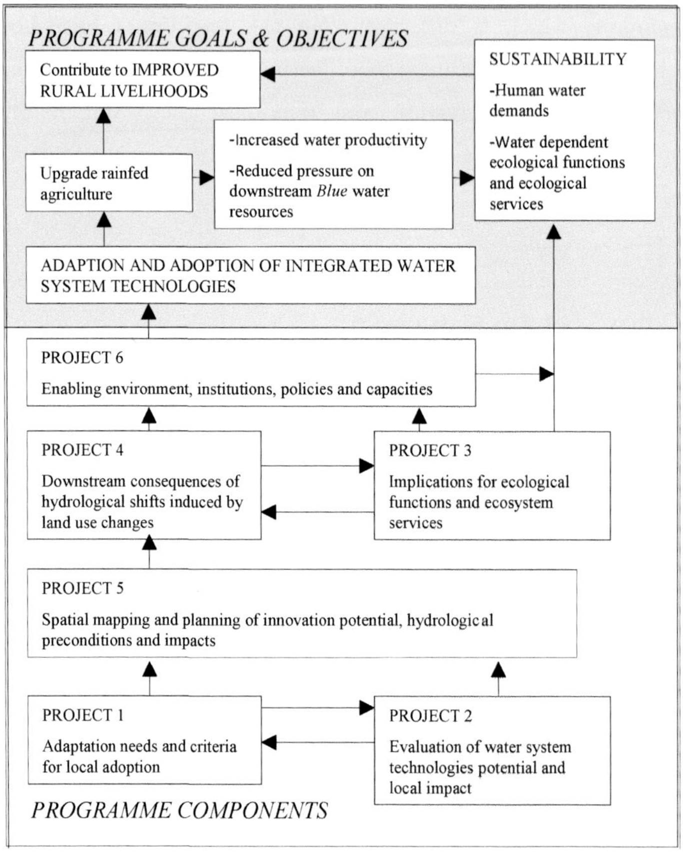

Figure 1.1 The interaction and integration off SSI projects (SSI, 2002).

1.3 THIS RESEARCH

The research reported in this thesis falls under Project 2 of the overall SSI programme. The research evaluates the existing agricultural water use technologies and compares them with introduced "more efficient" alternative cultivation techniques. In the process, water balance investigations have been conducted based on on-site observations to quantify water partitioning at field scale. A better understanding of partitioning processes at field scale leads to a better understanding of crop water productivity and how this can be enhanced from the little available water.

The research reported in this thesis shows the linkage with other projects in the programme and has allowed participant farmers to bring in their own ideas into the final output.

Structure of thesis

This thesis is presented in eight chapters as summarised below:

Chapter 2 presents the challenges faced by rainfed subsistence farmers and what options exist to break out of the food scarcity problems. This leads to the objective of the research.

Chapter 3 presents the site selected for the research, the selection of farmers to work with and the characteristics of the study area.

In Chapter 4 the setting of observation methods is presented with an explanation of the techniques which have been tested in the research. Findings from the on-site measurements are also presented.

The findings from on-site measurements lead to the application of modelling techniques to better understand the water balance components at the field scale. Different modelling approaches are presented which complement each other to explain the partitioning of the available water. This is presented in Chapter 5.

The partitioning analysis and yield observations lead to Chapter 6 which focuses on water productivity in rainfed agriculture.

Chapter 7 provides a synthesis of the whole research based on the setting of rainfed systems in sub-Saharan Africa, the current practices and scope for improvement. A reflection on linkages with other SSI projects is also presented.

Chapter 8 presents the conclusions of this research.

Chapter 2

Rainfed agriculture in sub-Saharan Africa

2.1 Water and food security

In 2000 the United Nations adopted eight goals to be achieved by 2015. The first of these Millennium Development Goals (MDGs) is the target to halve hunger by 2015 (UN, 2005). Unfortunately, as far as the first goal is concerned, not much ground appears to have been covered in sub-Saharan Africa (UN, 2008) and chances of achieving the goals by the set target are getting slimmer everyday. Sub-Saharan Africa contains the highest number of malnourished people and has the largest increase in the number of hungry people (FAO, 2001). The annual water use for crop productivity is 7 130 x 10^9 m^3 a^{-1} (de Fraiture et al., 2007) while the economically available water is 9 000 – 14 000 x 10^9 m^3 a^{-1} (FAO, 2003). This shows that the largest use sector of the available water is agriculture. The demand for water for food increases by 1300 m^3 ca^{-1} a^{-1} for each additional person (Rockström et al., 2009). While the competition for water continues to grow, the food demand, hence demand for water for agriculture, is projected to increase by up to 90% in 2050 (de Fraiture et al., 2007).

Water for all purposes is extracted from the hydrological cycle. Water availability within the hydrological cycle is better illustrated from the rainbow of water concept (Savenije, 2000).

Rainwater is converted to runoff in rivers and groundwater in aquifers, which drains to the oceans or lakes. This is defined as "blue water". The component of water that infiltrates into the soil to replenish the root zone is defined as "green water". From an agricultural perspective, transpiration from the green water stock is considered "productive" evaporation that is converted to biomass. Green water use for global crop production is significantly higher than consumptive blue water use (Hoff et al., 2009). Blue water is used in most day to day livelihood activities such as domestic use, commerce and industry, hydropower generation, recreation and irrigation. "White water" is the part of the rainfall that evaporates before it can infiltrate into the soil (through canopy and surface interception) and is considered as "unproductive" evaporation, as compared to the "productive" transpiration.

Figure 2.1 illustrates this rainbow of water concept.

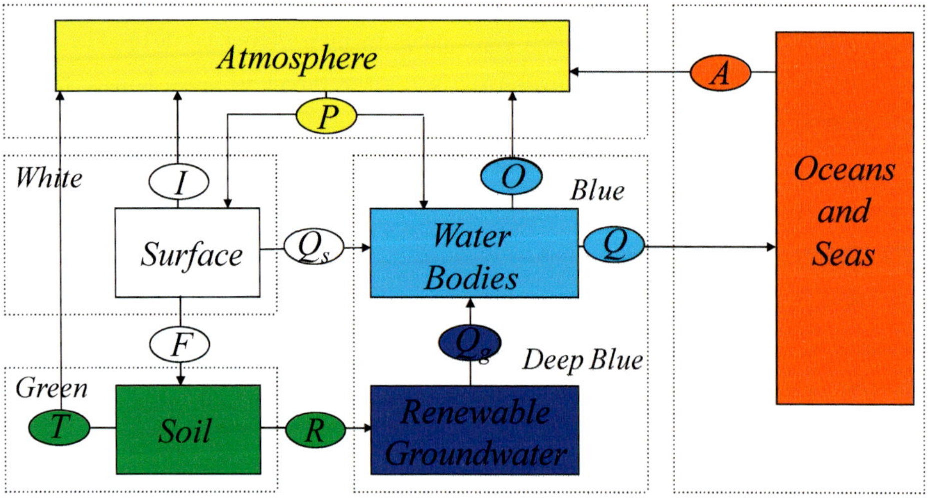

Figure 2.1 The rainbow of water as conceptualised by Savenije (2000).

Figure 2.1 does not show the "unproductive" soil evaporation that draws on the green water stock, but is also part of "unproductive" white water.

Attempts have been made to quantify the amounts of water allocated to various hydrological processes within a smallholder farming system. Rockström (2001), based on research in the semi-arid tropics in Africa, found that out of the rainfall received, 15-30% is taken up by transpiration processes (green water), 30-50% as interception and soil evaporation (white water) while runoff and surface storage accounts for 10-25% (blue water). Figure 2.2 shows that the balance (10-30%) is deep percolation which eventually recharges groundwater (deep blue water).

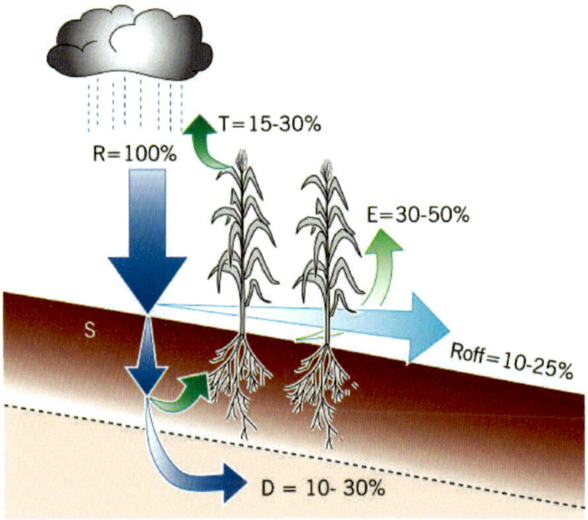

Figure 2.2 Partitioning of rainwater at smallholder farm scale (Rockström, 2001).

From this assessment it is clear that only a small proportion of rainfall is used productively from an agricultural perspective. More yields can be obtained if ways can be found to re-partition this water balance so that more water is converted to transpiration purposes. To achieve this, ways should be found to reduce direct evaporation and excess runoff hence diverting more water for biomass production through transpiration.

Competition for water has increased in the recent past and has been accelerated by recent global trends to treat the environment as a legitimate use sector. This competition has resulted in water scarcity particularly for the blue water sector. For a long time in the past, the solution to water scarcity has been the construction of additional infrastructure but, nowadays, economic considerations and resistance from environmental groups has seen less new infrastructure being built. Water scarcity is so critical in some basins to the extent that some are now "closing down" with no uncommitted flow left (Falkenmark and Molden, 2008; Mazvimavi et al., 2007). For the 30-year period up to 2000, irrigated areas have risen sharply in developed countries but, for developing countries particularly in Africa, no significant changes have been noted (FAO, 2002).

Water scarcity is not equal in all parts of the world (Kijne et al., 2009) with different parts of the world experiencing either physical or economic scarcity (Humphreys et al., 2008). In the context of food security, water scarcity should not be viewed as just a shortage of water for drinking or basic household requirements, but that it is more of lack of sufficient water to produce food for basic human survival at household level (Savenije, 1998). A global analysis of green and blue water availability suggests that water stress is primarily a blue water issue and large opportunities are still possible in the management of rainfed areas i.e. the green water resources (Rockström et al., 2009). As a result, blue water scarcity is increasing with new projects for blue water facing more resistance from other stakeholders and environmentalists. Consequently, there is now a decrease in funding for blue water related projects including dam construction and irrigation development. The challenge in semi-arid climates is that, whereas they receive comparable, if not more, annual rainfall than in temperate climates, the distribution of rainfall is not favourable and it results in more dry spell occurrences during cropping seasons (Barron et al., 2003). These dry spells severely impact on yield levels. Unless interventions are made, 59% of the world will face blue water shortage while 36% will face green and blue water shortage although of course this is largely depended on detailed patterns of change in climate, demography, land use and technical development (Rockström et al., 2009).

At a global scale, enough food can be produced to feed the world's population, yet, the number of under-nourished people increased from 840 million in 2002 to 923 million in 2007 (FAO, 2002; FAO, 2008). Hunger, poverty and water scarcity are strongly related (Rockström et al., 2007). This means that efforts to attain food security and reduce poverty should be strongly linked to water management, especially green water management since a large proportion of the poor and hungry relies on rainfed farming for their livelihoods. Any attempt to determine if there will be enough water to support food production for the projected world population of 8 billion by 2025 requires an understanding of the link between water availability and food production (FAO, 2002).

Figure 2.3 shows that global maize yield trends have increased over the last two decades but the trend in sub-Saharan Africa has not improved during the same period (FAO, 2005).

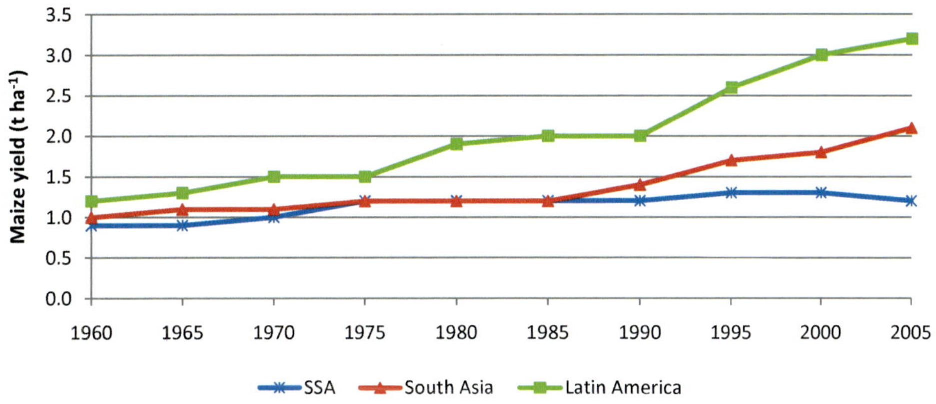

Figure 2.3 Maize productivity levels in selected sub-continents in the past decades (adapted from FAOSTAT, 2005).

2.2 Water productivity

Water productivity relates to the yield (biomass or grain) derived from using a specific quantity of water. For crops, a higher crop water productivity (CWP) results in either the same production from less water resources, or higher production from the same water resources (Zwart and Bastiaanssen, 2004). The factors which affect CWP include crop type, water availability and soil, agronomic and economic factors (Ali and Talukder, 2008). From the current consumption trends, it is projected that, unless if improvements in productivity are made, crop water consumption will increase by 70% - 90% by 2050 (de Fraiture et al., 2007). With the increasing threats for water scarcity, there is the choice of allocating more blue water to support required food levels or, alternatively, improve productivity in rainfed systems, so that up to 2 200 x 10^9 m^3 can be freed to other uses (Kijne et al., 2009).

Agricultural droughts and dry spells pose the biggest challenge in improving yields for smallholder farming systems. Agricultural droughts occur for periods between 3-5 weeks during the growing season and can damage the final crop performance if they are not properly managed (Rockström, 2003a). Dry spells, on the other hand, last for periods of 10-21 days within the growing season. Maize, for instance, which is the staple food in most sub-Saharan African countries, is highly sensitive to water stress and requires a gross rainfall of 500-800 mm/season to mature while sorghum, which is a more drought tolerant crop requires a gross rainfall of 450-650 mm/season (Critchley and Siegert, 1991). This means therefore that if all such crops were entirely rainfed, there would be just sufficient water to support staple food crops in the region (Hatibu, 2002).

While the shortage of water is believed to be a key limiting factor to crop production for many agricultural systems, it can also be argued that the impact of dry spells can be significantly reduced as, in practice, water use efficiencies are generally low (Makurira et al., 2007a). Hence, even in semi-arid conditions, water productivity can be increased if more of the available water can be channelled to productive purposes. Productive use, in the context of this thesis, is defined as that use which promotes transpiration and, hence, biomass production.

The water available to a plant as green water is much less than what may be believed to be the case. Mean rainfall estimates do not give a proper indication of water availability as dry spells during seasons affect the performance of a crop even if the mean rainfall received is about the average expected value. Depending on the crop growth stage, water shortage at crucial stages may damage a crop significantly (Barron et al., 2003). It is therefore important to explore ways of increasing CWP under the given challenges.

Dry spells significantly impact on productivity levels. In Tanzania, for instance, drought induced famines occur during 33% of the time (Hatibu, 2002). Such meteorological droughts are difficult to manage and usually result in total crop failure

(Rockström et al., 2007). In general, mid-season dry spell occurrences are more frequent. It has been observed that rainfall variability within seasons plays a more influential role in crop production than total seasonal rainfall occurrences. Maize, for instance, is sensitive to moisture stresses at particular growth stages and yields are more affected by water availability at critical growth stages than by the total seasonal water availability (Barron et al., 2003). This means therefore that the water scarcity, from an agricultural perspective, for semi-arid environments is often a result of agricultural droughts where dry spell impacts especially at critical growth stages affect the resultant yields at the end of the season (Barron et al., 2003).

Where the threat of water scarcity is high, a shift to less water demanding crops can be the solution to guaranteed harvests. Crops perform differently to moisture stress variations as shown in Table 2.1. This means that, in areas which dry spells occur more frequently, crops with lower sensitivity to moisture stress should be grown.

Table 2.1 Sensitivity of common food crops in SSA to moisture stress (adapted from Critchley and Siegert (1991).

Sensitivity scale	Crop
Low	groundnuts
↓	sorghum
	cotton
	sunflower
	beans
High	maize

The amount of rainfall received at a particular area is only an indication of the potential of such an area to support crop production. Actual water availability to support crop growth is a function of water partitioning at the given point. In this type of analysis other factors such as soil type, management of soil nutrients and overall farm management practices are usually ignored yet they contribute significantly to overall crop performance (Kijne et al., 2009). However, for large parts of rainfed agriculture, variable rainfall and subsequent crop water availability is the major constraint causing low yields with poor water productivity. A better understanding of water partitioning at plot scale hence serves to promote better management of dry spells.

2.3 WATER FOR FOOD IN SUB-SAHARAN AFRICA

Hatibu (2002) states that the total renewable water resources in sub-Saharan Africa is 4000 x 10^9 m^3 a^{-1} and this water is sufficient for food security in the region if only it were to be available at the right place and at the right time. Within the year, rainfall variability is very high with, for instance, yield reductions being experienced three years out of five in some parts of Zimbabwe due to rainfall variability (Mugabe et al., 2002). Total crop failure due to droughts is experienced once in every ten years in semi-arid sub-Saharan Africa (Ngigi, 2003). Kenya, for instance, receives between 150 and 1200 mm a^{-1} yet the average potential evaporation is between 1200 - 3000 mm a^{-1} (Mbugua and Nissen-Petersen, 1995) thus making effective rainfall insufficient for basic crop production unless if interventions are made to alter this partitioning.

However, crop productivity is low in sub-Saharan Africa where yields are lower than global trends. In rainfed sub-Saharan Africa, the potential yield of maize is about 6 t ha^{-1} yet the average is only 1.4 t ha^{-1} (de Fraiture et al., 2007) with many drier parts realising harvests of less than 1 t ha^{-1} (Bhatt et al., 2006; Rockström et al., 2004).

The situation in sub-Saharan Africa therefore requires greater attention if food security is to be improved. Attention towards irrigation schemes is not a long term solution as the number of beneficiaries in irrigated areas is much smaller than that of rainfed farmers. Regionally, 90% of the food production originates from rainfed areas which means that improving productivity on existing irrigated areas will only have small impacts on the general food supply (de Fraiture et al., 2007).

Sub-Saharan Africa is therefore faced with the challenge of feeding the highest population who, at the moment, consist of the largest number of undernourished and most poverty stricken in the world. The region is generally classified as facing economic water scarcity (Falkenmark, 1995; Humphreys et al., 2008). This means that there is more water that can be used but what lacks is the means to access the water. With 90% of the local food production coming from rainfed agriculture (de Fraiture et al., 2007) and, acknowledging that water scarcity is rather a blue water issue than green water, attention should focus more on improving CWP in rainfed agriculture if any significant mileage is to be covered towards the attainment of the MDGs.

2.4 SCOPE FOR IMPROVED CROP PRODUCTIVITY

Improved productivity can be achieved from a number of perspectives. More land can be opened up for agriculture from where more food can be produced. Alternatively, more blue water can be diverted to irrigation purposes to minimise water stress in growing seasons. In reality, these options are not very feasible as most of the productive land in many countries is being utilised anyway. As discussed before, it is not easy to divert more water to irrigation as blue water scarcity is already being felt in many parts of the world. Given that 90% of food production in sub-Saharan Africa is met from rainfed areas any efforts towards irrigation facilities will often result in insignificant incremental yields (de Fraiture et al., 2007). The third option is that of increasing the productivity of rainfed farming systems. This option is more promising as it involves the conversion of unproductive water processes to more productive "green water" processes.

The scope to improve CWP exists (Kosgei et al., 2007; Mupangwa et al., 2006; Mwenge Kahinda et al., 2007). The current yield levels in rainfed systems which average 1 t ha^{-1} are well below the potential levels of 6.6 t ha^{-1} (de Fraiture et al., 2007). This implies that, if this yield gap could be narrowed, substantial ground can be covered in achieving food security at smallholder farming scales. More than 80% of African countries have the potential to double their CWP (Liu et al., 2008).

Kijne et al. (2009) rightfully point out that the greatest potential to increase yields lies in the areas where productivity levels are still low at the moment. Sub-Saharan rainfed agricultural systems are a perfect example of such areas as, except for a few cases, the region relies entirely on rainfed agriculture (Hoff et al., 2009).

Increased productivity in rainfed systems can be realised in several ways. The rainfall patterns in SSA are characterised by high intensity and short duration events with almost instantaneous but short-lived runoff events. Rainfall events do not last up to an hour and rainfall intensities in excess of 30 mm hr^{-1} have been observed in the study area during the research period. The rainfall variability is very high with 70% of it being lost to unproductive purposes (Hatibu, 2002). In such situations, rainwater harvesting is appropriate to capture such rainfall for use in drier periods (Ngigi, 2003; Pacey and Cullis, 1986; Reij et al., 2001) recommend the application of rainwater harvesting techniques for rainfall above 150 mm a^{-1} but also emphasise that the benefits are more evident for rainfall figures of about 500 mm a^{-1}. Hatibu et al. (2006) acknowledge that rainwater harvesting exhibits more benefits in average to above average seasons and no significant impacts in poor seasons.

Besides rainwater harvesting, other soil and water conservation techniques can be applied to improved water productivity in rainfed systems. These include conservation tillage (Hatibu, 2002; Rockström et al., 2001). These methods minimise soil disturbance (hence reduce direct soil evaporation) while promoting more infiltration and root development. Proper soil and water conservation efforts allow for a win-win situation where more water is availed to the root zone while, at the same time, soil and nutrient loss are minimised. In addition to reducing soil erosion, these conservation efforts also help to reduce flood hazards.

With more focussed investments in rainfed systems, water savings of 15 – 20% can be realised in the next decade (Hoff et al., 2009).

2.5 RESEARCH AND INFORMATION GAPS

It has been discussed that a lot of research has been conducted worldwide on assessing water availability to support food requirements for the present and, also, for the projected future. There is overwhelming agreement that, although there may be enough water to produce global food requirements, the distribution of rainfall results in water scarcity for agriculture in some parts of the world. In areas where such scarcity occurs, e.g. sub-Saharan Africa, hunger, poverty and malnourishment are common and chances of achieving the MDGs are slim as major interventions are required to overturn the current negative trends.

It seems clear that issues of water for food require urgent attention. It is equally clear that the solution to the challenges of water for food security should start with the promotion of soil and water conservation in rainfed farming systems.

However, there is not much evidence to suggest that substantial ground has been covered in translating this theory into practice. There is a big information gap on comprehensive studies to prove that, indeed, there is scope to more than double the present yields in rainfed farming systems under the same environmental and climatic conditions. In the absence of such comprehensive and convincing research, the adoption of promising farming practices will take too long to realise.

Jewitt (2006) notes that water storage in the soil profile and its partitioning to evaporation from the soil, transpiration, groundwater recharge and to different parts of the downstream flow regime is the least understood aspect of the hydrological cycle. Yet this is the process which defines crop water productivity.

The big challenge, therefore, is to gain more insight into on-farm soil moisture dynamics to produce food under harsh climatic conditions. It is equally important to understand the implications of up-scaling water system innovations at field scale to the wider catchment and basin scales. The promotion of more efficient techniques at field scale generally implies that more water is diverted to infiltration and, hence, transpiration processes while reducing soil evaporation, deep percolation and surface runoff generation. While positive impacts may be realised at local field scales, there are also concerns that the adoption of "more efficient" systems at field scale will result in more water being converted to biomass production processes which implies less water availability for downstream ecosystems. Again, this is an information gap which needs to be filled.

Scientifically, it is important to understand water balances at field scale. However, establishing water balances and moisture transitions at field scale is complex given the high variability of terrain, cropping patterns and high temporal and spatial variability of water distribution at micro scales. On the other hand, it is also important to understand this field scale hydrology if efficient water use activities at field scale are to be promoted and up-scaled for the benefit of many. Increased efficiency at field scale implies increased use of water at micro level for transpiration purposes while less efficient water use at field scale is associated with less infiltration, higher evaporation and more runoff, i.e. more water to components of the hydrological cycle other than green water. Upgrading rainfed agriculture, in other words, implies maximising green water flows from available rainfall and the runoff generated at micro-scale. Fears exist that this improvement in efficient use upstream compromises water availability at wider downstream scales.

The information gaps identified here need to be closed through obtaining a better understanding of the processes that are involved. This means that more data should be collected through a wide range of scientific approaches which range from in-field observations to larger spatial data collection techniques using, e.g. remote sensing and GIS techniques. When more data is available, it is also possible to predict, with better precision, the impact of upscaling any local successes to basin-wide scales.

2.6 OBJECTIVES AND RESEARCH QUESTIONS

2.6.1 Main objectives

The objective of this research is to contribute towards food security in rainfed systems by evaluating the efficiency of the current farming practices, and to propose and test the effectiveness of alternative and more efficient water use techniques.

This objective is realised through a better understanding of soil water processes and moisture dynamics at field scale.

This objective has been achieved by meeting the following specific objectives:

i. To evaluate the efficiency levels of crop productivity under the current and alternative farming practices;

ii. To measure the potential impacts of improved farming systems under scarce water environments;

iii. To conduct water balance analyses at field scale based on on-site observations and participatory experimentation; and

iv. To predict potential improvements in productivity levels based on re-partitioning of the field water balances.

2.6.2 Key research questions

i. Which agricultural water management practices are currently being employed at smallholder level in the study area? [Chapter 3]

ii. What are the shortcomings in the current farming practices at smallholder scale? [Chapter 3]

iii. What is the scope for introducing innovations to improve biomass production? [Chapter 4]

iv. What are the hydrological implications of key water system innovations in relation to rainwater partitioning, green water efficiency and overall rainwater use efficiency at field scale? [Chapter 5]

v. What are the implications of the introduced techniques on crop productivity under the current hydro-climatological challenges? [Chapter 6].

Chapter 3

THE STUDY AREA[2]

3.1 PHYSIOGRAPHY

The research has been conducted in the Makanya catchment in Northern Tanzania (Figure 3.1). The Makanya River is a tributary of the Pangani River which originates from the Kilimanjaro region and drains into the Indian Ocean covering a catchment area of about 42 000 km^2 (Pangani Basin Water Office, 2005). The Makanya River drains a catchment area of 320 km^2 (Enfors, 2009). The headwaters of the Makanya River originate in the South Pare Mountains at an altitude of about 2100 m (Mul et al., 2008) and flows through subsistence rainfed farming systems before discharging into the Pangani. The terrain changes dramatically from mountainous in the upper part of the catchment (Vudee villages) to midlands (Bangalala and Mwembe villages) and, finally, flatter lowlands at the downstream end of the catchment (Makanya village).

---

[2] Based on the following papers:

a) Makurira, H., Mul, M.L., Vyagusa, N.F., Uhlenbrook, S. and Savenije, H.H.G., 2007a. Evaluation of community-driven smallholder irrigation in dryland South Pare Mountains, Tanzania: A case study of Manoo micro dam. Physics and Chemistry of the Earth 32(15-18): 1090-1097.

b) Mutiro, J., Makurira, H., Senzanje, A. and Mul, M.L., 2006. Water productivity analysis for smallholder rainfed systems: A case study of Makanya catchment, Tanzania. Physics and Chemistry of the Earth, 31(15-16): 901-909.

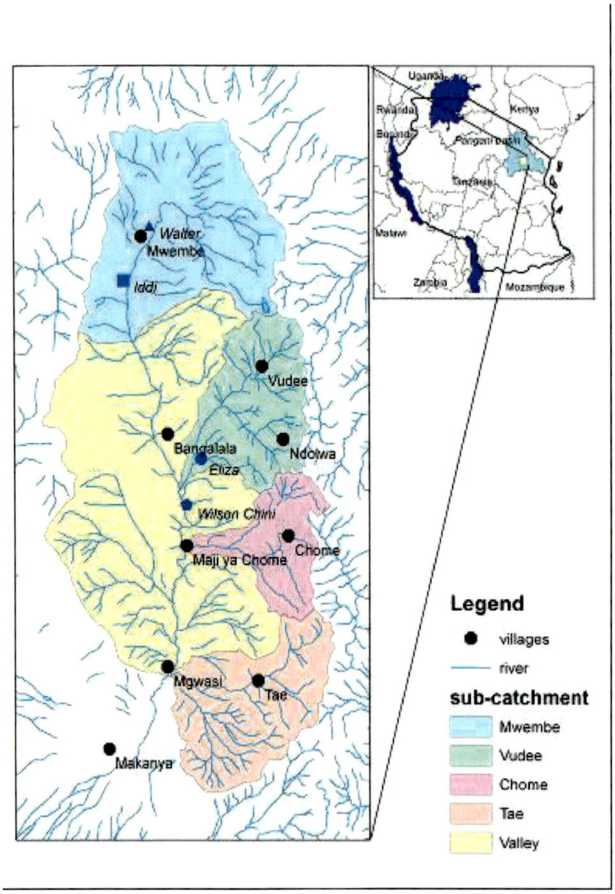

Figure 3.1 The Makanya Catchment.

Four tributaries, Mwembe, Vudee, Chome and Tae, join to form the main stream in the Makanya catchment. The Makanya tributary discharges into the Pangani River only on a few occasions in a season when exceptional floods are realised in the upper part of the catchment which cannot be absorbed by upstream activities and processes. Community elders tell of stories of how the flow regime used to be perennial up to the late 1970s but has since become seasonal.

3.2 RAINFALL

The Makanya catchment receives rainfall ranging between 400 – 1000 mm a^{-1} depending on altitude and season (Enfors et al., 2008). The highlands receive generally more rainfall than the lower parts. The region, being close to the equator, has a bimodal type of rainfall distribution with a short season occurring anytime between October and December (locally known as *Vuli*). The longer season (locally known as *Masika*) occurs between March and June. This means that the annual rainfall received is split over two agricultural seasons which indicates that there is hardly enough water to support the common food crops such as maize and beans (Mutiro et al., 2006). The rainfall type is that of short duration and high intensity which has a rapid runoff response, but only for short periods. This runoff, if not harvested, drains into the river networks and alluvial aquifers (Mul et al., 2007) before reaching occasionally the outlet of the catchment. An analysis of the rainfall pattern for the nearby Same meteorological station for the period 1957-2004 reveals a steady mean in the total annual rainfall received but an increasing trend in dry spells of 21 days or more has been observed for the *Masika* season (Enfors and Gordon, 2007).

Figure 3.2 shows an analysis of rainfall data for the period 1934-2007 at Same meteorological station by Mul (2009). The analysis concluded that there are no statistically significant trends in the cumulative short season, long season and annual rainfall records. A visual analysis, however, suggests an increasing trend in the total seasonal rainfall in the short seasons *(Vuli)* with a declining trend in the long seasons. The annual pattern is showing a declining trend as well which would imply that the long rainfall seasons dominate the overall annual pattern.

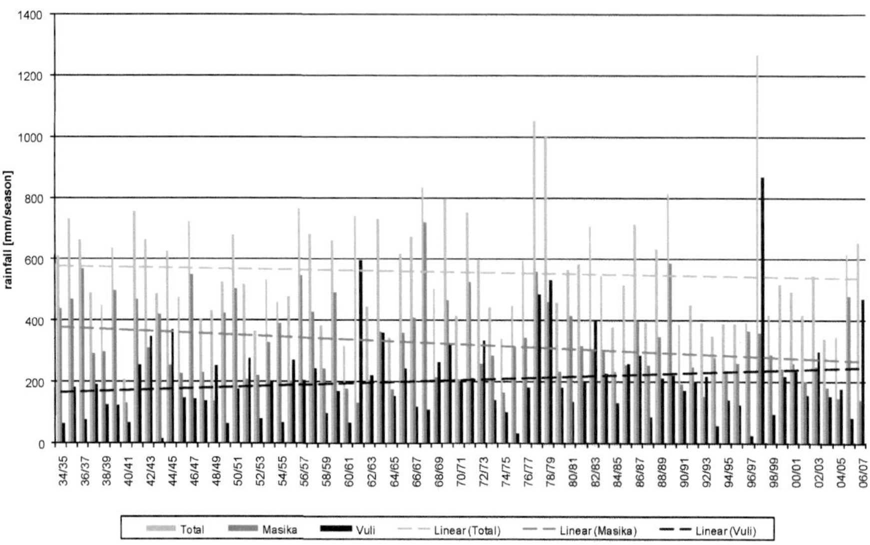

Figure 3.2 Rainfall trends at nearby Same station for the period 1934-2007 (source: Mul, 2009).

3.3 Demography

The population in the Makanya catchment is estimated around 35000 and is increasing at an estimated growth rate of 1.6 % per annum (URT, 2004). Taking into account that 90% of the population in the Pare Mountains live in the highlands, of which 80% depends directly or indirectly on agriculture (Mwamfupe, 2002), innovations for improved agricultural productivity out of a diminishing water resource base are necessary. Settlement patterns suggest a strong influence of migration from the mountains into the midland areas as a result of increasing population densities (Fischer, 2008). However, this migration from the highland areas implies a shift from wetter climate to drier environments where the natural resource base does not offer many comfortable livelihood options.

Rural poverty is of concern to the general Tanzanian population and the Government has taken measures to alleviate this situation (URT, 2005). In Maswa District, Hatibu et al. (2006) report that the lower quartile of the poor is very poor with a mean expenditure of only US$ 0.1 per capita per day. The mean of the upper quartile is reported to be US$ 0.9 per capita per day. The Government of Tanzania has already acknowledged the need to prioritise rural poverty by, among others, promoting agricultural production for most food and cash crops especially to smallholder farmers (URT, 2005). Animals are also kept as part of the wealth at household level but these are only disposed of in extreme hunger situations (Enfors and Gordon, 2008).

3.4 SOILS

The soils in the study area vary between sandy, clayey and loam (c.f. Enfors et al., 2010). The variation in soil texture is very high. The highlands and midlands are characterised by shallow soils with mainly granite bedrock. The soils in the lowlands are mainly alluvium deposits from the upstream erosion processes. The alluvium soils have higher moisture holding capacities and contain fine sediment deposits which are rich in nutrients from upstream agricultural activities. This partly explains why decent harvests can be obtained in the lowlands even with only two or three flood events which facilitate spate irrigation.

3.5 CURRENT LAND USE PRACTICES

The communities in the research area are predominantly subsistence farmers who derive their livelihoods from local land and water resources. The increasing population has exerted excessive pressure on the land in Makanya, especially in the highlands where people were originally settled. As a result, some people opted to migrate to the midlands and lowlands which are, basically, drier but more sparse (Fischer, 2008). The most dominant land use is crop production and livestock rearing.

In normal years, the community relies mostly on agricultural production from the two rainfall seasons. In extreme dry years, reliance on local production systems can diminish from 80% to 20% (Enfors and Gordon, 2008) depending on the severity of the dry spells. In such cases livelihoods cannot be sustained by agriculture alone, hence they resort to alternative coping mechanisms such as marketing available food stocks (including livestock) or relying on food supplements from working relatives in cities.

A variety of cultivation techniques is observed within the study area. The most common cultivation technique is the use of the hand-hoe. After the wetting of the soil surface from early rains, the hand hoe is used to break the soil crust after which planting takes place in shallow hand dug pits. Only a few farmers use the ox-drawn plough and this is mainly attributed to lack of resources to procure or rent implements. On steeper slopes, cultivation is conducted on terraces. These terraces are reported to have been constructed through support from a non-government organisation which offered incentives for constructing terraces. Terracing can therefore be described as a new farming practice in the area.

3.6 WATER FOR FOOD SECURITY AND COPING STRATEGIES

The highlands receive comparatively more rainfall than the rest of the catchment with average annual rainfall of about 1000 mm a^{-1}. In such wetter environments, the communities are engaged in different agricultural practices which include grain crops, fruit trees animal, rearing and fisheries. Traditional farming practices are mainly driven by climate characteristics (Rockström et al., 2007). In these wetter upland areas where water scarcity is less, the farming activities involve comparably higher water demanding crops. Crops in the highlands hardly experience water stress. The midland and lowland areas receive rainfall of about 400 – 600 mm a^{-1} and 300 – 400 mm a^{-1}, respectively. This means that these areas do not generate sufficient rainfall to support livelihoods hence the farmers have adopted other rainwater harvesting techniques such as runoff diversions, micro dams (*ndivas*), earth excavations (*lambos*), tie-ridges and terraces (Rockström et al., 2004). Micro dams are very popular as they store water for supplemental irrigation. The number of micro dams has been increasing due to an increasing need for extra water to mitigate dry spells. The construction of micro dams is favoured because micro dams are cheaper to construct and require relatively less technical expertise compared to larger storage structures.

The resultant indigenous irrigation schemes in the South Pare Mountains, which have been in existence for a century now, are communally managed with external support only obtained at setting up the schemes. The idea of community managed irrigation scheme is very interesting especially given the high value attached to water resources particularly towards dry spell mitigation. Generally, communities group together and identify a potential site for a micro dam and, in some fortunate instances, assistance to construct or rehabilitate such micro dams is obtained from donors while the members provide the rest of the inputs, including labour. At first sight, these micro dams are ideal for small irrigation blocks but, in line with traditional norms and values, new players are not often shut out. In practice, the accommodation of an infinite number of participating members compromises the volumes available for allocation to each member.

Even with a network of more than 100 diversion canals and 75 micro dams (Mul et al., submitted) the small size of these micro dams (usually less than 200 m^3) and the large command areas result in inefficient utilisation of the diverted water (Makurira et al., 2007a) and, consequently, do not offer effective solutions for dry spell mitigation.. Due to the large number of these micro dams in the Vudee sub-catchment, water is shared between upstream and downstream users and, also, among schemes within the same part of the catchment. Under a local agreement between the upstream Vudee village and the downstream Bangalala village, upstream users abstract water during the day and leave night flows for downstream users. The other upstream village, Ndolwa, only releases excess water. Between the two micro dams in Bangalala village, diversions to specific micro dams occur on three specific days of the week. Abstractions on Sundays are not regulated. Normally, water fills up the downstream micro dams at night and irrigation occurs during the day (Mul et al., submitted). This is a good example of hydro-solidarity (Kemerink et al., 2009), which has the potential to avoid conflicts if it is fully implemented.

In the absence of solid agreements for sharing water between the upstream and downstream stakeholders, the lowland areas of Makanya catchment rely only on excess flood flows for spate irrigation. Locals claim that the Makanya river used to be perennial up to the 1970s. Since then, only extreme flood events in the upstream catchment may result in some flow being available at the lower end of the catchment.

In an average season, only one or two such flood flows are experienced at the bottom of the catchment. However, the high water holding capacity of the soils usually means that this water, when diverted onto the fields, is sufficient to support crops to maturity in the lowlands.

The Makanya catchment can therefore be described as characterised by soil degradation, nutrient loss and a general shortage of water for agricultural purposes. Given the low seasonal rainfall, insufficient river flows for supplementary irrigation and the traditional practices which favour relatively high water demanding crops such as maize, more efficient cultivation techniques are required to cushion the farmers against these challenges.

3.7 YIELDS OBTAINED

Maize and bean varieties are the most common food crops grown in the study area (Kimaro et al., 2009; Mutiro et al., 2006). Maize, being the staple food crop is the most popular with maize harvests being the most common measure of how good a season has been or how good a farmer is. Most of the farmers can be classified as subsistence farmers with not much excess incomes to go beyond basic household requirements. As a result, farming activities are also affected by lack of surplus cash to buy treated seeds, fertilizers, pesticides or more modern farming implements. The seed used is basically selected from previous harvests and, being untreated, may result in diminished yields at the end of the season.

The average annual rainfall received in the area is 600 mm a^{-1} and is very low for CWP given that this rainfall is split into two growing seasons. Because of this low rainfall and the dilution of the potential impacts of supplemental irrigation facilities, the grain yields obtained are very low and oscillate below 1 t ha^{-1} (Bhatt et al., 2006; Rockström et al., 2004). This poses a serious threat to food security given the increasing population levels and the challenges to meet the MDGs. Water scarcity for agriculture is believed to be the major reason why the yields obtained for the maize crop are very poor for farmers who do not practice soil and water conservation techniques.

Therefore, there is a need to at least double the current yield levels if projected food demands are to be satisfied from local sources.

3.8 TRADITIONAL FARMING PREFERENCES

Traditional preferences present a challenge to researchers and policy makers. The scenario described in this section of low and highly variable rainfall potential, high evaporation rates, threats of exceedance of carrying capacities of soils and a general lack of resources to purchase appropriate seed, fertiliser, pesticides and herbicides. This combination of unsuitable conditions does not favour the cultivation of sensitive but preferred crops like maize. Instead, less water demanding crops should be explored. Consideration of virtual trade is another possibility since, in general, the farmers prefer to store their wealth in the form of livestock.

Many studies that have been undertaken usually have findings which, if pursued, will result in notable improvements. In reality only a few findings are adopted. This is partly attributed to the gap between the research and the beneficiaries of such research. Modern research now seeks to involve beneficiaries during the period of research. This participatory research approach has higher chances of adoption than the previous approaches (Mirghani and Savenije, 1995). Interestingly, even where participatory research has been conducted, adoption is not always automatic even where research successes are very clear (Bewket and Sterk, 2002; Mirghani and Savenije, 1995).

A challenge therefore exists to explore what influences adoption. This challenge has been taken up within the SSI programme. One reason could be that there is a gap between policy pushes and general research. Policy makers need to be convinced beyond doubt of research successes before they can be convinced to change policy. This is easier to achieve if participatory tools are used at research stage (c.f. Chikozho 2005; Swatuk and Motsholapheko, 2008; Tumbo et al., 2010).

3.9 DISCUSSION AND CONCLUSIONS

This chapter has described the study area. The characteristics of this site are similar to many other smallholder rainfed farming systems in SSA where the challenges of infertile soils, over-population, and lack of resources to invest in more efficient farming practices and the threat of dry spells all contribute towards sub-optimal yields.

Adoption to more efficient farming systems such as less water demanding crops or virtual water trade are examples of ways of escaping from this vicious cycle that keeps local populations in the poverty trap. For instance, the farmers can grow beans and sorghum which are less water demanding, market them and use the income to buy staple food. Alternatively, they can sell their livestock to buy food. However, traditional values play a key role in shifts from current practices. It is therefore necessary that research and policy directives be targeted at the vulnerable communities to convince them of the need to adopt more efficient practices.

This research is one of many efforts to contribute towards the improvement of livelihoods in vulnerable systems. A participatory approach has been adopted for this research whereby farmers obtain an opportunity to understand the research objectives and contribute to the research through experimental setting, data collection, and, hopefully, analyse and appreciate the findings from the research.

Chapter 4

RESEARCH SITES AND OBSERVATION TECHNIQUES[3]

4.1 SITE SELECTION

Four sites within the Makanya catchment have been selected for detailed research. These sites are owned and managed by four different farmers using similar traditional farming practices. The study sites are shown in Figure 3.1 as Wilson (Site 1), Eliza (Site 2), Iddi (Site 3) and Walter (Site 4). The selection of the sites has been random as the research could only take place where the owner expressed willingness to participate in the research. Furthermore, the existince of a runoff generation area upstream of the study site and a flow channel to collect the generated runoff also influenced the final site selection. A participatory research approach has been adopted where the farmer managed his/her farm as much as possible with the researcher designing the experiment and the farmer participating in the monitoring

---

[3] Based on the following papers:

a) Makurira, H., Savenije, H.H.G. and Uhlenbrook, S., 2007b. Towards a better understanding of water partitioning processes for improved smallholder rainfed agricultural systems: A case study of Makanya catchment, Tanzania. Physics and Chemistry of the Earth, 32(15-18): 1082-1089.

b) Makurira, H., Savenije, H.H.G., Uhlenbrook, S., Rockström, J. and Senzanje, A., 2009b. Investigating the water balance of on-farm techniques for improved crop productivity in rainfed systems: A case study of Makanya catchment, Tanzania. Physics and Chemistry of the Earth, 34: 93-98.

and interpretation of the data. Since the four sites are spread over a spatial distance of 10 km, site characteristics such as soil properties, slope and daily weather patterns can differ substantially. The research seeks to compare the performance of the maize crop at the different sites under similar treatments.

At all four sites it has been observed that, although some forms of improved farming techniques have been observed in the study area, these techniques are not widespread. This is typical of many farming systems in the study area and the rest of the SSA region. The only significant exceptions are the terraced areas where construction has been heavily influenced by external agencies (e.g. NGOs). In the absence of soil and water conservation efforts the overworked soils and rapidly degrading soils are exposed to severe erosion which normally is associated with uncontrolled nutrient loss. The soil and nutrients are washed away into drainage channels where they mostly contribute to the siltation of rivers and are mostly deposited in downstream lowland plains.

The soils in the study area can be described as sandy to loamy sand. Limestone occurs sporadically with reddish soils in the uplands which turn yellowish in the lower parts with alluvium deposits found in the river beds (Mul, 2009). According to Mutiro et al. (2006), the soils in the research area are of an alkaline nature (pH range 7.4 – 8.8). The geology is characterised by igneous rocks which have undergone regional metamorphism (Mul et al., 2007). The drainage is defined by fault lines that intersect the catchment (Mul, 2009).

4.2 BACKGROUND TO THE DEVELOPMENT OF TESTED TECHNIQUES

The research has tested alternative farming practices which have the potential to demonstrate that more efficient farming practices exist and which, if adopted, can result in significant changes in general livelihoods through increased crop productivity. Consideration has been given to promising researches which have been concluded under similar challenging environmental and climatic conditions as experienced in the study area (Agarwal et al., 2005; Barron and Rockström, 2003; Ngigi et al., 2007; Rockström et al., 2001). The research prioritised simple and affordable solutions which are more likely to be taken up by the community after the expiry of the research period.

The research has also acknowledged that it may not be feasible to identify and test the "best" solution to the challenges faced. Different solutions exist for different conditions. Rather, the research seeks to demonstrate that, indeed, there is scope to improve on the currently obtained yield levels if only more efficient practices are adopted. The advanced "innovations" have been tested to check their impact on improving maize yields while, in the process, providing a better understanding of hydrological processes at field scale.

In choosing the techniques to be tested in the research, it has been noted that traditional farming systems are associated with the use of implements such as the plough and hand hoe. These implements are believed to lead to the formation of a hard crust at the soil surface. This crust is usually broken at the onset of the rainfall season using available implements such as ploughs and hand hoes. The breaking of this crust accelerates the rate of soil moisture loss due to evaporation. On the other hand, the use of the hand hoe and the plough promotes the formation of a hard pan at shallow rooting depths. Infiltration potential is therefore highly compromised while, the hard pan formed from hand hoe use, interferes with root development to deeper depths. This type of cultivation is inefficient from a soil moisture retention perspective (Rockström et al., 2001).

It is also acknowledged that most rural farmers lack resources to invest in improved farming systems hence are not likely to adopt techniques which require large investment costs. If the farmers could have surplus cash, they could possibly purchase treated and more appropriate seed varieties. They could also consider applying commercial fertilisers or invest in irrigation infrastructure. Better yields are more guaranteed with larger and focused investments as shown in Figure 4.1.

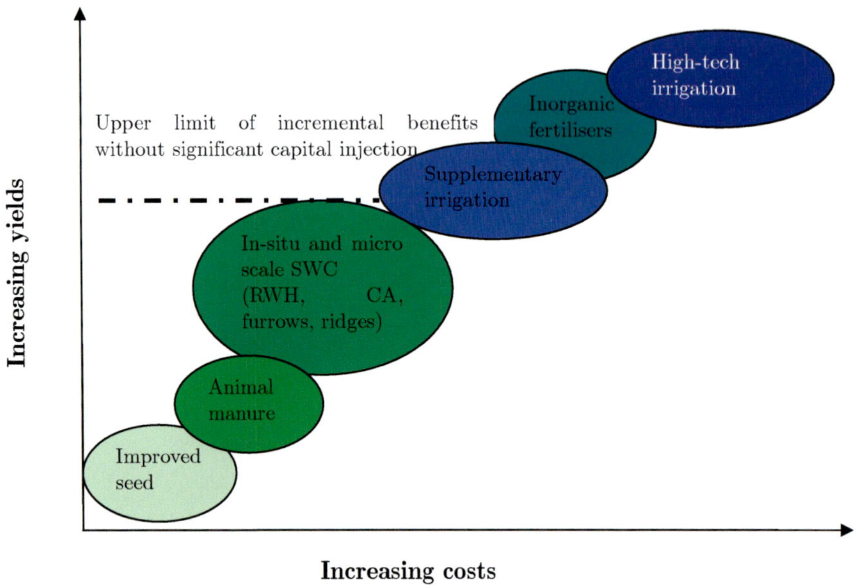

Figure 4.1 A schematic illustration of options for improving agricultural water management in dryland cropping systems (adapted from Nyagumbo et al., 2009).

The research has therefore taken place against a background of little seasonal rainfall, frequent dry spell occurrences, during and in between cropping seasons, and ineffective soil and water harvesting. This research builds on the hypothesis that positive impacts in land and water management and, ultimately, improved water productivity should take into consideration the following aspects:

4.2.1 Conservation tillage

The Magoye ripper has been advanced as a substitute for the traditional plough and the hand hoe. The ripper is animal drawn and pierces through the soil making thin and sharp lines along intended planting lines. The ripper can penetrate depths of up to 40 cm and cuts through the hard pan. This means that only the planting line is opened thus promoting infiltration along the planting line while, at the same time, minimising soil evaporation at the soil crust. These lines are maintained in subsequent planting seasons.

While the ripper is time and labour efficient, the major disadvantage of this technique is that it does not clear weeds hence weed management is important especially during the first few seasons of adopting this technique (Rockström et al., 2001). The use of agro-chemicals for weed control is out of the question in the study area as it is unaffordable for the majority of the farmers. The ripper requires draught power which is not always available like the common hand-hoe.

The soil and water conservation method considered for further research in the study area is *fanya juu* cultivation technique. *Fanya juus* are a soil and water conservation technique practiced in some parts of East Africa and Central Africa (Bewket and Sterk, 2002; Nyagumbo et al., 2009; Tenge et al., 2005). They are basically soil ridges within cultivated land where trenches are dug across the field and the excavated soil is placed upslope to form bunds within the plot.

Generally, the *fanya juu* technique is applied where steep slopes, hence, greater potential for runoff generation exist. High runoff rates are also associated with rapid soil and nutrient loss. The *fanya juus* hence serve a dual purpose of both soil and water conservation although only their soil conservation functioning has been widely publicised (Gichuki, 2000; Tenge et al., 2005). Besides conserving soil, the second, and more important function from a water conservation point of view, is that the runoff generated is forced to fill the constructed in-field trenches before the excess water overflows downslope along the cultivated area and finally drains out of the field. The trenches therefore act as temporary water storages which allow infiltration to occur for longer periods during and after the flood and diversion events. The reduction in runoff velocities results in the deposition of fine sediments in the trenches and along the bunds across the field. This deposition of fine sediments also results in nutrient enrichment of the soil within the cultivated field and, most likely, alters the soil water holding capacity.

The *fanya juu* technique has been successfully tested in East and Southern Africa especially with regards to soil and water conservation and increased resultant yields (Gichuki, 2000; Motsi et al., 2004; Mwangi et al., 2001; Tenge et al., 2005). While these successes have been explained from a soil conservation point of view, the other dimension, which is more important from a water productivity point of view, the hydrological functioning of these in-field structures has not been explored in detail. The understanding of their hydrological functioning provides better insights into the impact of these interventions on yield levels. Immediate questions which come into mind are:

- What are downstream impacts of the success of these interventions in runoff generating areas?
- What are the impacts on hydrological processes at entire catchment scales?
- Do these interventions reduce floods and soil erosion?
- Do these interventions contribute towards groundwater recharge?

Figure 4.2 shows how the infiltration potential is more enhanced in the trenches and at the ponding zones as a result of increased residence time of water.

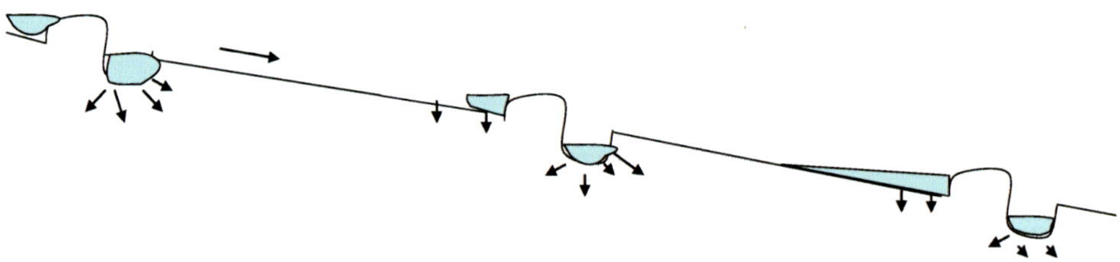

Figure 4.2 The *fanya juu* cultivation technique showing zones of increased infiltration potential.

While it is acknowledged that conservation agriculture may not be the panacea for the agricultural productivity challenges experienced in SSA (Giller et al., 2009), it however, offers scope for stability in obtained in communities where conditions are not favourable to attain close to decent yields from season to season. According to Pacey and Cullis (1986), rainwater harvesting (and possibly conservation agriculture as well) is only viable in areas where the minimum rainfall is 200 mm a^{-1}. The Makanya catchment fits in this category. If the rainfall is excessive, the impacts of rainwater harvesting become insignificant.

4.2.2 Seed selection

The study area is characterised by short rainfall seasons dominated by frequent mid-season dry spells. Seed selection should therefore focus on short season varieties which are more drought resistant. The Kito maize seed variety developed in Tanzania has been chosen as the best type of seed for research. The variety is a 90-day crop with better tolerance to dry spells and, hence, offers more secure harvests. The disadvantage, however, is that the cobs and grains are smaller compared to the longer season varieties. From a food security point of view, the drought resistant seed varieties enhance the chances of, at least, obtaining a harvest even during average and slightly below normal rainfall seasons.

4.2.3 Cover cropping and manure

Cover cropping is also practiced in the study area with a combination of maize and beans being the most common. This is practiced mainly to enhance food security options and has no consideration to nutrient cycles or moisture competition. Practically, this technique helps to reduce evaporation while improving nitrogen cycles in the soil. The research adopted this practice for testing it in combination with other soil and water harvesting techniques.

Since the farmers are generally resource poor, the majority of them cannot afford to buy the commercial fertilisers. Because of this disadvantaged position, the rainfed farmers now resort to the application of animal manure as fertiliser. Animal manure is a viable substitute to commercial fertilisers provided the nutrient levels of the applied manure are high enough. This cannot be guaranteed in the overgrazed and degraded areas like the research site. In this research, manure has been applied at a rate of 5 t ha^{-1}. This application rate has been adopted following recommendations from Selian Soil Research Institute in Arusha who conducted extensive soil analyses to ascertain the general classification of the soils and their nutrient levels.

4.2.4 Rainwater harvesting

Rainwater harvesting is common but is not practiced efficiently in the study areas. A number of rainwater harvesting techniques can be observed across the entire Makanya catchment ranging from terracing, micro storage dams and diversions. The majority of farmers within the study area rely on supplemental irrigation from micro dam storages. However, these schemes have proved to be inefficient under the current management practices hence their benefit can be described as insignificant (Makurira et al., 2007a).

In this research we tested diversions from runoff generated in gullies and other informal flow paths close to the study fields. These flows have been concentrated onto the research site and measured. This type of rainwater harvesting is not practised in the study area to date and, if proven successful, could be one of the most important innovations to influence the water balance at farm level. Because additional water can only be provided when it rains, the innovation essentially aims at more efficient partitioning of rainfall and enhancing moisture storage and availability in the root zone.

4.3 Typical experimental setting

Figure 4.3 shows the typical experimental set-up at each site. A control plot has been taken upslope of the *fanya juu* terraces so as to eliminate the influence of runoff diversion. The control plot is therefore entirely rainfed. In the control plot maize under strict rainfed farming condition has been grown using the hand-hoe technique as is the current practice. For the experiment, the trials have been conducted within a *fanya juu* set-up with access to runoff diversion. Runoff has been diverted from gully flow as this has been identified as a major source of local runoff generation and is not commonly utilised by smallholder farmers where it is available. Site selection has, therefore, been also influenced by the presence of a gully in the vicinity which has the potential to collect runoff before discharging into main drainage channels. The sizes of these gullies, similar to upstream catchment areas, can vary between sites. At each site, the cultivated plot (between the *fanya juu* trench and bunds) has been divided into four sections with a different treatment for each section. Under Treatment 1 the effect of exclusive ripping with runoff diversions into the trenches and the field plot has been tested, while in Treatment 2 the effect of ripping in combination with manure applied at a rate of 5 t ha^{-1} every season and access to rainwater diversion have been tested. For Treatment 3 the effect of the traditional hand hoe tillage technique in combination with manure application and access to rainwater harvesting were tested, while at Treatment 4 the effect of using the hand hoe tillage only with access to diverted water has been tested. The treatments are explained in Table 4.1.

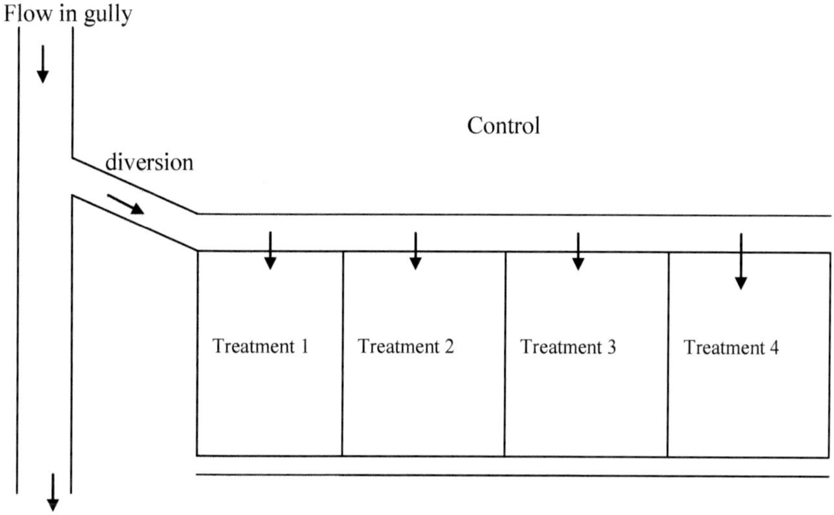

Figure 4.3 Typical experimental set-up showing the location of the treatment plots in relation to the diverted water.

Table 4.1 Description of treatments.

Treatment	Description
Treatment 1	Exclusive ripping with rainwater harvesting.
Treatment 2	Ripping in combination with cover cropping and manure applied at 5 t ha^{-1} and access to rainwater harvesting.
Treatment 3	Traditional hand hoe tillage technique in combination with manure application and access to rainwater harvesting.
Treatment 4	Hand hoe tillage with access to harvested water.
Control	The common cultivation practice by using the hand hoe under strict rainfed systems.

Since the individual farmers have been allowed to manage their farms as far as possible, the level of adoption of the advanced techniques differed between farmers. For instance, at Site 1 and Site 2, the *fanya juus* have not been constructed according to recommendations as, apparently, the respective farmers were not convinced of the potential of the trials. On the other hand, the farmers at Site 3 and Site 4 the farmers followed the researcher's recommendations more comprehensively. Figure 4.4 shows a typical experimental field after a rainfall event.

(a) (b)

Figure 4.4 *Fanya juu* trenches and rip lines across a cultivated strip showing (a) water stored in the trenches after a rainfall event, and (b) much wetter rip lines after a rainfall event at the onset of a cropping season.

4.4 CONCEPTUAL FRAMEWORK

The main objective of this research is to investigate if there is scope to increase productivity levels in smallholder farming systems under the existing hydro-climatic challenges of water for agriculture. To achieve this objective, it is also necessary to better understand water partitioning processes at field scale. Only through a better understanding of the hydrological cycle at field scale can it be determined how much of the available water is being attributed to productive purposes (green water fluxes). Thereafter, it would be possible to propose and evaluate more efficient farming techniques which aim to optimize these "green water" fluxes at the expense of non-productive fluxes. This way it would be easier to identify promising farming options which can be presented to farmers for adoption.

Complementary methods have been adopted in this research to evaluate the effectiveness of advanced techniques from hydrological, agronomic and soil science perspectives.

The following water balance equation provides the conceptual framework for a better understanding of partitioning processes at field scale:

$$\frac{dS}{dt} = P - E - Q$$ **Equation 4.1**

where (all terms in mm d^{-1}),

> $\frac{dS}{dt}$ is the rate of change of storage in the system,
> P is the precipitation received in the system,
> E is the evaporation from the system, and
> Q is the outflow from the system measured as runoff.

4.5 PARAMETERS MEASURED

Basic parameters (rainfall, runoff diversion, yields) were measured at all the four sites. However, because the farmers at Site 3 and Site 4 adopted the research in more detail, limited additional instrumentation (e.g. lysimeter, runoff generation) were prioritised at these sites. Also, since the farmers adopted the recommendations of the research more comprehensively it is believed that these sites provide a more accurate measure of the interventions that have been tested.

4.5.1 Rainfall

Precipitation *(P)* has been measured using a locally fabricated rain gauge but following the World Meteorological Organisation (WMO) guidelines for a standard raingauge of diameter 12.5 cm and placed at least 30cm above the ground. Daily rainfall records have been collected at 9 a.m. every day.

4.5.2 Net in-field runoff contribution *(Q_s)*

Runoff has been directed onto the experimental plots through a controlled inlet point and has been allowed out through one exit point. Tipping bucket loggers have been installed to measure surface flow into and out of the study plot. The difference between inflow and outflow is the net surface flow contribution, Q_s.

4.5.3 Soil evaporation *(E_s)*

Soil evaporation can be measured by direct or indirect methods:

(a) Direct methods by lysimeter measurements

Balancing weights

A lysimeter has been constructed at Site 4 based on the principle of balancing moments. Two drums, one with a representative soil sample and another with concrete as counterweight constitute the lysimeter setup. The side with the representative soil has a known cross-sectional area (A_d). After a rainfall event, this weight increases by an amount equal to the weight of water captured in the drum. The evaporation is measured as the reduction in weight from the moist soil in the drum at 9.00 a.m. every day. The difference in weight for each day is attributed to evaporation from the lysimeter. Figure 4.5 shows the lysimeter that was fabricated on site for soil evaporation analysis.

Figure 4.5 The lysimeter set up at Site 3 to monitor soil evaporation.

The evaporation from the lysimeter at each time step (E_L) is determined from the relationship

$$E_L = P - \frac{\Delta W_E - \rho \Delta V_L}{\rho A_d \Delta t}$$ **Equation 4.2**

where,

E_L	daily evaporation from lysimeter (m d^{-1});
P	daily rainfall received (m d^{-1});
ΔV_L	volume of water added directly into the drum (m^3);
ΔW_E	manual weight added (kg);
ρ	density of water (kg m^{-3});
A_d	cross sectional area of drum (m^2); and
Δt	time step (d).

Automatic strain gauge

A strain gauge has been installed in the setup described above and it has been equipped with a data logger. The strain gauge therefore recorded continuous weight differences, ΔW, at set intervals, Δt.

If the strain gauge data is analysed for a fixed daily time, then Equation 4.3 can be used to determine daily evaporation from the lysimeter (E_L).

$$E_L = P - \frac{\Delta W - \rho \Delta V}{\rho A_d \Delta t}$$ **Equation 4.3**

where,

P	daily rainfall received (m d^{-1});
ΔV	volume of water added directly into the drum (m^3);
ΔW	change in weight over observed period (kg);
ρ	density of water (kg m^{-3});
A_d	cross sectional area of drum (m^2); and
Δt	time step (d).

(b) Indirect methods by soil moisture observations and yield measurements

Indirect methods involve the calculation of soil evaporation from the soil moisture and climatic observations. These are described in more detail in Chapter 5.

4.5.4 Soil moisture measurements

To evaluate the effect of different treatment techniques, soil moisture has been measured by use of the time domain reflectometry (TDR) technique. Four access tubes have been inserted into the ground using a hand auger. Tube A has been placed in the control plot, i.e. the section within the same research field but upstream of the diversions. This section is strictly rainfed. The other three tubes have been placed within the same cultivated strip (and between the *fanya juu* trenches) but located such that Tube B is closest to the trench, Tube C is in the middle of the cultivated strip and Tube D at the lower end of the strip and closest to the bunds. From Figure 4.6 it can be seen that Tube B monitors the impact to the root zone of water stored in the trench while Tube D monitors the effect of ponded water as a result of the bund. Tube C monitors the soil moisture in the middle of the plot.

For the subsequent downstream cultivated strips between the *fanya juu* constructions, TDR tubes have been placed only at the centre of field (corresponding to the position of Tube C). Soil moisture has been observed twice a week during the rainfall season and once in two weeks during the dry season.

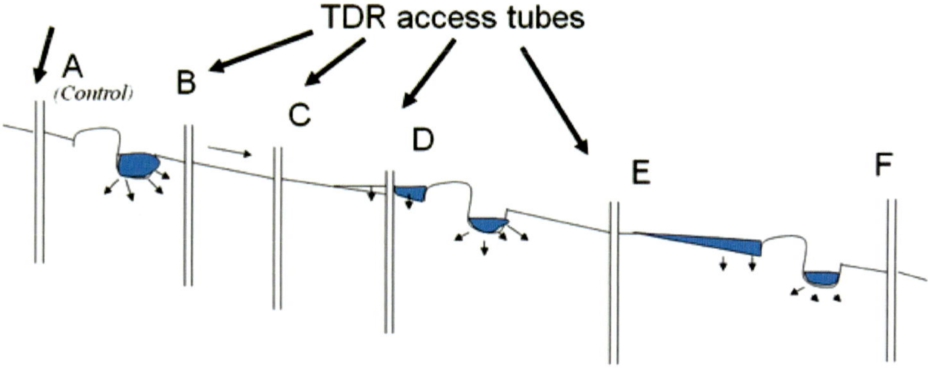

Figure 4.6 The set-up of the *fanya juu* technique with TDR soil moisture monitoring tubes.

4.5.5 Biomass and leaf area measurements

The Decagon Acupar meter (Decagon Devices Inc., 2004) has been used to measure the leaf area index of the growing maize crop. The equipment measures above canopy and below canopy radiation from which it calculates the leaf area index. Measurements have been taken at all sites on a weekly basis during the *Masika* season of 2006.

4.5.6 Yield observations

The obtained yields for each treatment have been calculated by measuring the dry weight of grains harvested from each treatment. In the final season of the research, the grain yield for Treatment 1 has been measured along each rip line to investigate if there is variation in the yield obtained with distance from the *fanya juu* trenches. The data obtained has been analysed to obtain grain yield (kg ha^{-1}) at each site and for each treatment.

4.6 RESULTS

4.6.1 Rainfall

The rainfall observed at each research site confirms the general understanding of the dry conditions in the study area. Figure 4.7 shows seasonal rainfall variability at each research site. The variability is high. Some sites received almost double the amount received at other sites. However, the total seasonal rainfall received only exceeded 400mm/season in only one year during the research period (i.e. during *Masika* and *Vuli* of 2006). This confirms that the gross seasonal rainfall received is hardly enough to support crop growth especially when maize, the preferred food crop, is considered.

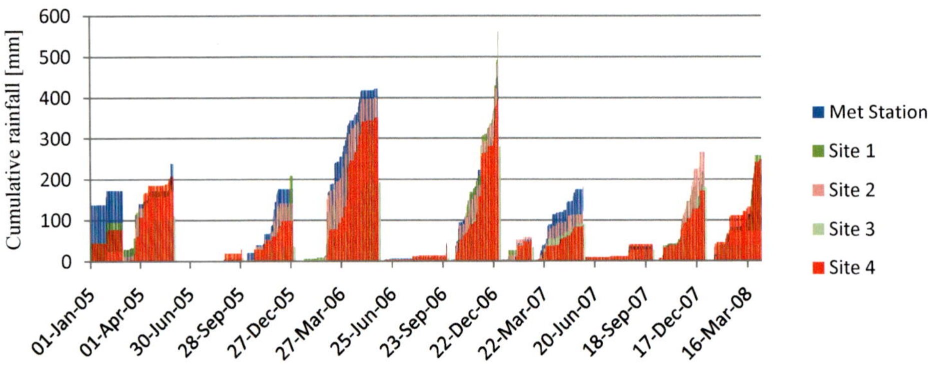

Figure 4.7 Cumulative seasonal rainfall observed at each research site between January 2005 and April 2008.

For cropping operations, the distribution of rainfall within the season is equally important as the seasonal totals. Dry spell occurrences especially during critical growth stages have been know to affect yield levels (Barron et al., 2003) even if seasonal totals may be good. At Site 4, for instance, dry spells of 10 days, or more, were observed at least twice in each season (except for both seasons in 2006). However, if it is considered that 2 mm d^{-1} is also lost to interception, then the frequency of dry spells can easily be four or five in a season which then impacts negatively on productivity.

Figure 4.8 shows the daily rainfall received at all sites for the long rainy seasons (*Masika*) (2005-2007). Again it is clear that rain days are not many and, except for a few exceptions, the rainfall received hardly exceeds 20 mm d^{-1}. The 2006 seasons experienced better rainfall distribution with a few good peaks although the season was comparably short. The 2007 season hardly experienced any rainfall events above 20 mm d^{-1}.

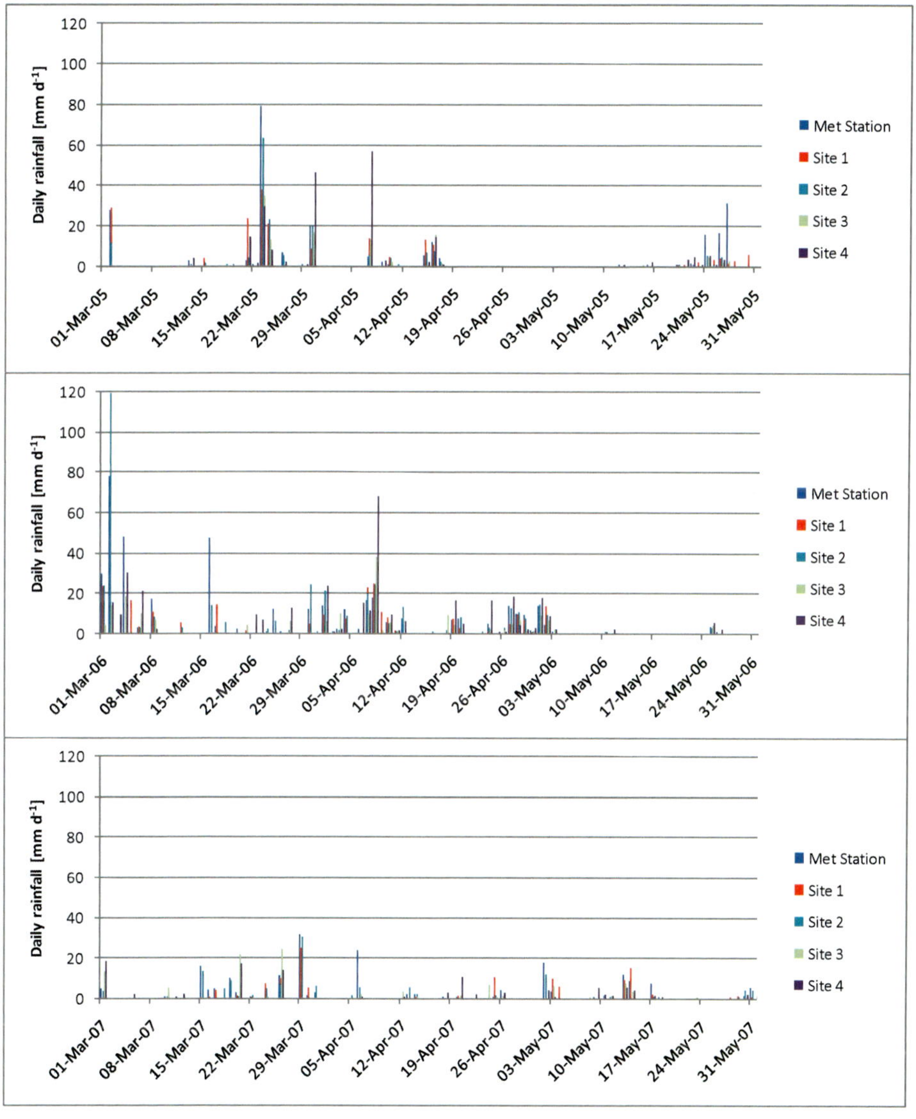

Figure 4.8 Daily rainfall received at the research sites during the long rainfall seasons (*Masika*) 2005 -2007.

Figure 4.9 magnifies the variation in daily rainfall received during the short rainy seasons, *(Vuli)* (2005-2007) at the research sites. Peaks of more than 20 mm d^{-1} are observed in 2006. The rainfall distribution is considered to be good for cropping seasons with less dry spell occurrences. The 2007 short season was characterised by low rainfall peaks throughout and dry spells of 10 days or more.

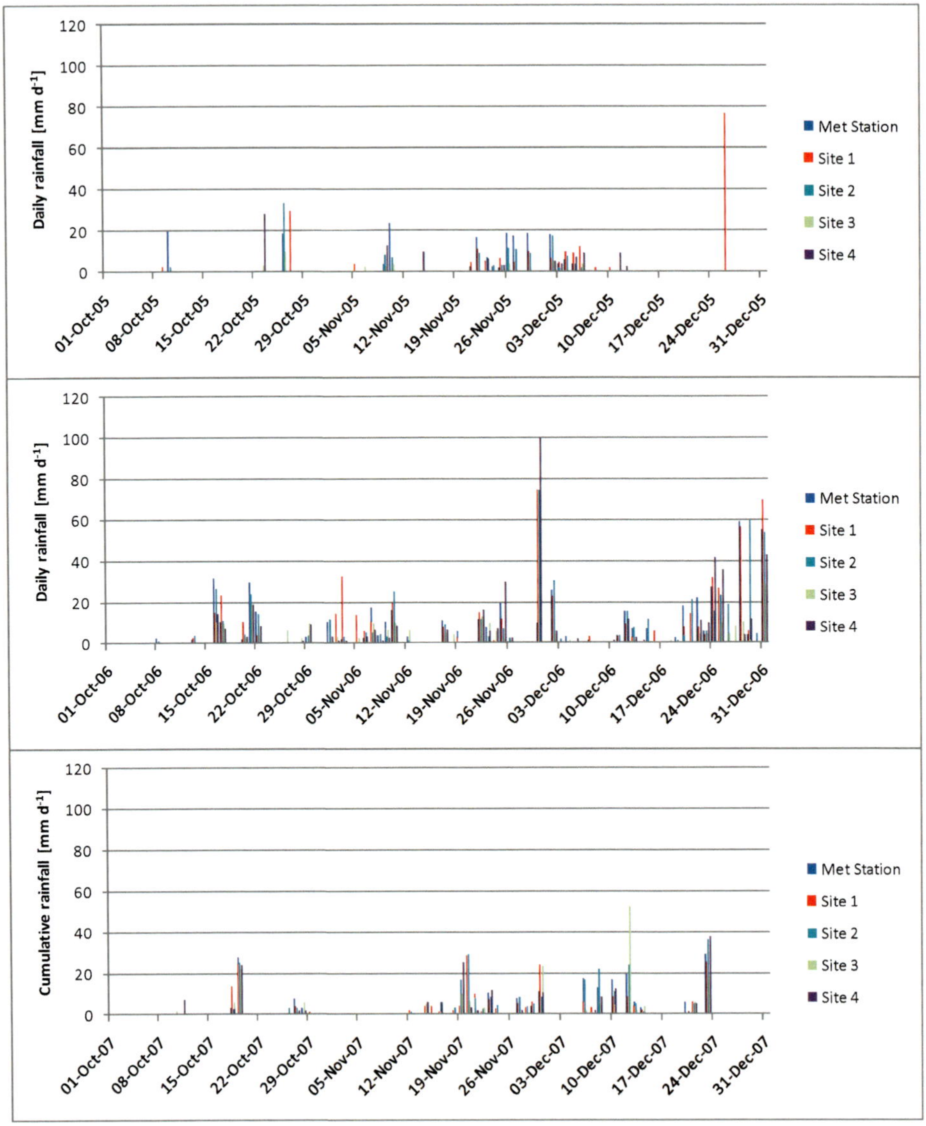

Figure 4.9 Daily rainfall distribution during the short seasons *(Vuli)* between 2005-2007.

4.6.2 Runoff generation

Short duration and high intensity rainfall events on, usually, crusted soil surfaces are normally associated with the generation of immediate runoff (c.f. Dlamini et al., submitted; van der Giesen et al., 2000). In Season 2 at Site 4, for instance, out of the 437 mm received during the season, 77 mm was received over two days with intensities of 31 mm hr^{-1}, 22 mm hr^{-1} and 24 mm hr^{-1}. This rainfall generates immediate runoff which is also of short duration and is largely dependent on the characteristics of the catchment area contributing to the runoff and the rainfall intensity. Figure 4.10 shows typical runoff hydrographs displaying the runoff generated from the upstream catchment at Site 4. Typically, the runoff generated lasts for only up to one hour.

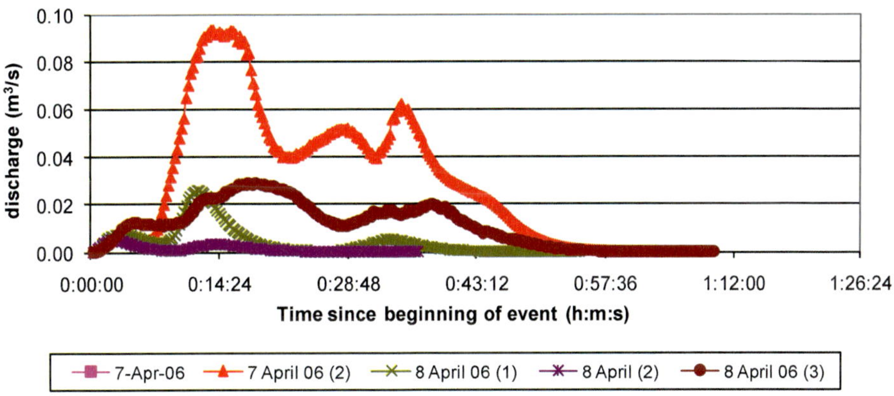

Figure 4.10 Typical runoff hydrographs at Site 4 at selected days in the long rainfall season (2006).

4.6.3 Net runoff contribution

Some of the runoff generated has been diverted onto the cultivated field to supplement the rainfall received. Figure 4.11 shows the cumulative rainfall and runoff diverted to Site 3 (Iddi) and Site 4 (Walter) between 2006 and 2008. Whereas Site 4 received comparatively more rainfall than Site 3, more water has been diverted to Site 3. Site 3 diverts more water than the actual rainfall received on site while Site 4 only diverts less than half of the rainfall received. The observed differences are mainly due to the difference in catchment areas.

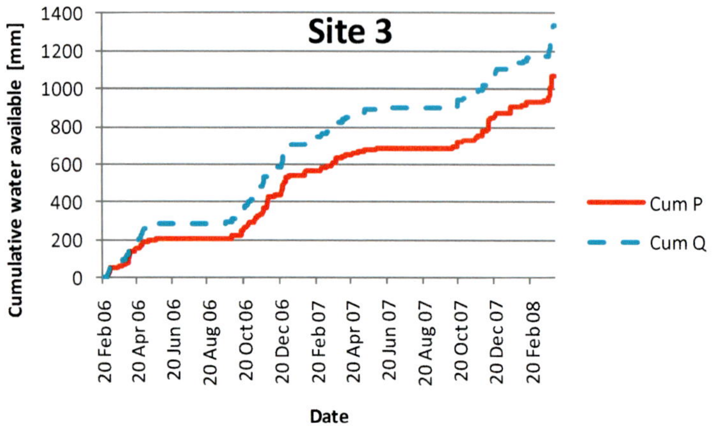

(a)

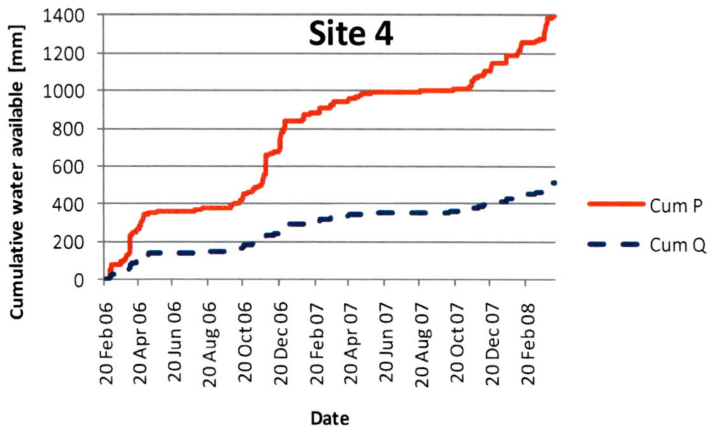

(b)

Figure 4.11 Comparison of cumulative rainfall (P) received and cumulative runoff diversion (Q) onto the field plots at (a) Site 3 and (b) Site 4 during the periods 2006-2008.

4.6.4 Soil moisture observations

Soil moisture has been monitored along a downslope transect as shown in Figure 4.4. Figure 4.12 shows the variation in moisture distribution during different seasons. The observations show a general trend where the tube closest to the bund (Tube D) records the highest soil moisture levels. On the other hand, the control section (Tube

A), generally records the lowest moisture levels. This shows that the diversions are effective in increasing soil moisture availability. An analysis of the three tubes between the *fanya juu* constructions (Tubes B, C and D) shows that Tube D records the highest moisture levels while Tube C records the least moisture levels. Tube B records levels which are somewhere between those of Tubes C and D. This pattern is more distinct at Site 3 than at Site 4 and can be attributed to the fact that more infiltration occurs at Tube D due to ponding effects. Tube B benefits from the stored water in the trench which may flow in a lateral direction. Tube C, being in the middle, benefits the least from these two processes. At Site 4, however, the steeper slope and shallower depth implies that water flows in a more lateral direction hence the moisture distribution does not show this pattern as clearly.

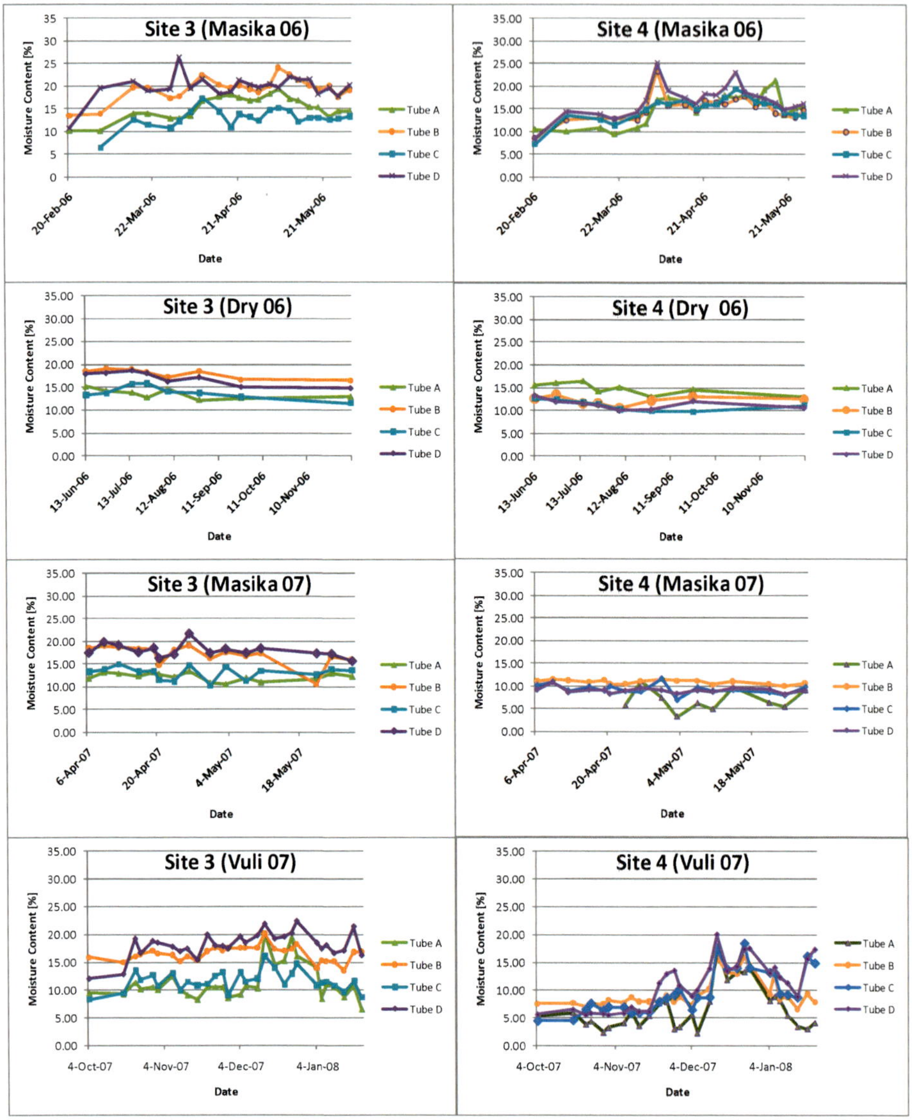

Figure 4.12 Soil moisture variations at the different locations of the tubes during different seasons.

<u>Statistical analysis of soil moisture variation</u>

Based on the soil moisture data collected, a statistical analysis has been conducted to investigate if there is significant difference between the soil moisture recorded at the different locations and the observed soil moisture in the control section. Figure 4.13

and Figure 4.14 show box plots of soil moisture variations along the longitudinal section throughout the observed period including both wet and dry seasons.

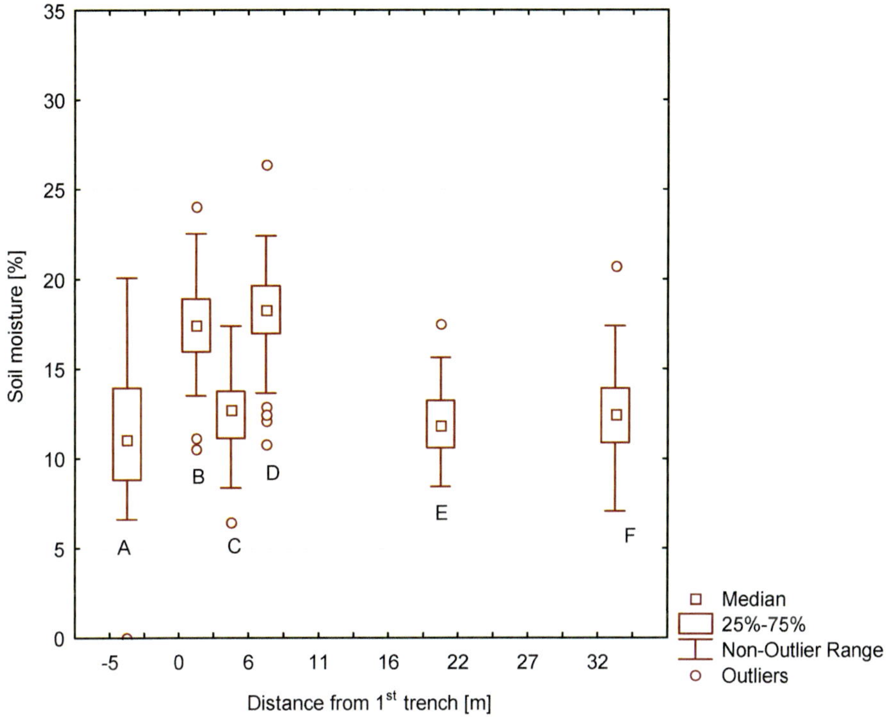

Figure 4.13 Soil moisture variations at different tube locations at Site 3 during all seasons.

Figure 4.12 shows that, at Site 3, the tubes between the *fanya juus* (B, C, D) benefit from the diversions, hence, record higher moisture levels. The tubes located at the centre of the field plots (e.g. Tubes C, E, F) do not show much variation in median moisture levels when compared with the control.

In Figure 4.14 the moisture distribution pattern is different with a gradual increase in moisture level with distance down the slope. Again, this suggests a more lateral soil moisture flux at Site 4 which is associated with the steeper slope and shallower soil depth compared to Site 3.

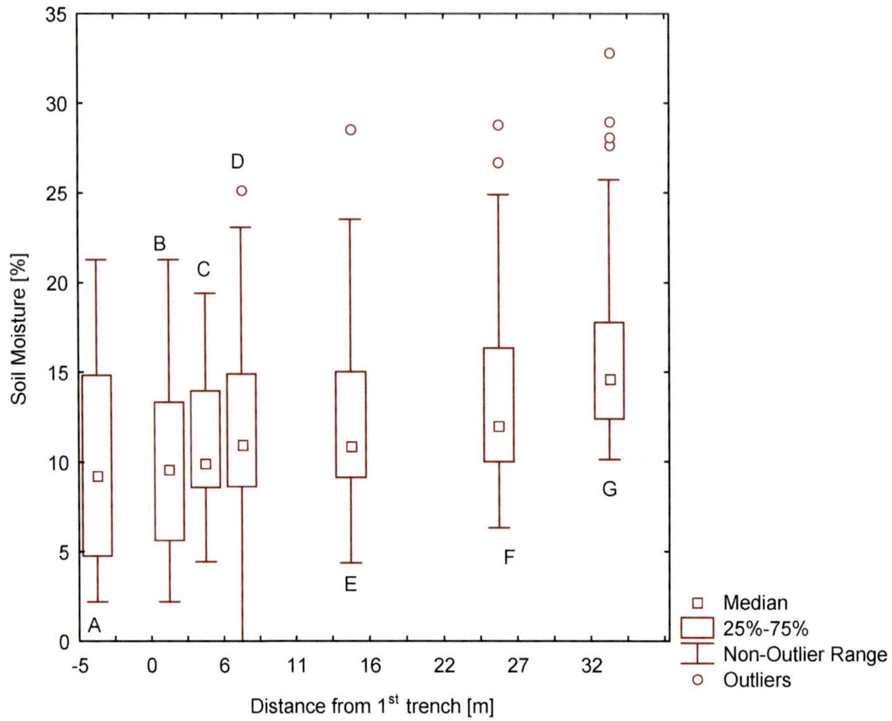

Figure 4.14 Soil moisture variations at different tube locations at Site 4 during all seasons.

4.6.5 Soil evaporation

Soil evaporation has been measured at Site 4 using an on-site fabricated lysimeter as described above. Observations have been made manually and, also, automatically using a strain gauge equipped with a logger. The manual observations form a longer series compared to the strain gauge readings. Figure 4.15 shows the cumulative evaporation from the manual readings and automatic recordings. A similar trend is observed between the manual and automatic graphs. The average evaporation rates recorded manually and automatically are 2.2 mm d^{-1} and 2.1 mm d^{-1}, respectively.

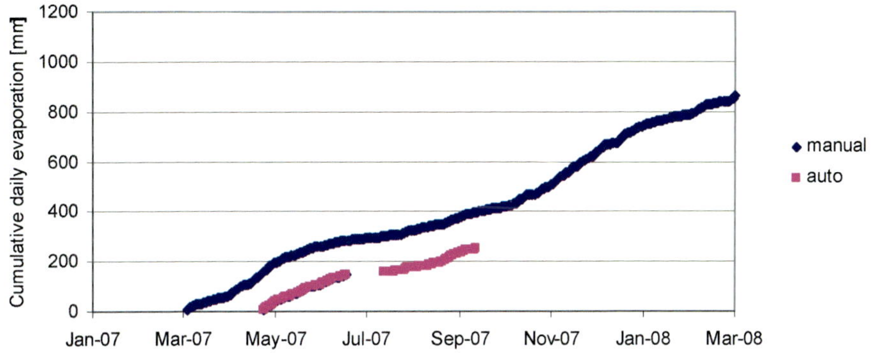

Figure 4.15 Comparison of daily soil evaporation rates measured manually and by use of an automatic strain gauge in 2007.

A comparison of evaporation data observed manually and automatically between January and February 2008 shows that the average evaporation by manual methods was 1.7 mm d^{-1} and 2.0 mm d^{-1} for the automatic method. The fact that the measurements from both methods are so well correlated gives confidence in the data collected manually. Manual methods of data collection, although at a daily time step, are simple and more sustainable.

4.6.6 Biomass measurements

The leaf area index at different stages of growth is plotted in Figure 4.16. It is observed that the leaf development is insignificant within the first two weeks of crop growth. Thereafter, the leaf area index becomes significant and increases up to about day 60 into the season after which it starts to fall at post maturity stage. The maximum leaf area index recorded here is 1.3 (m^2 m^{-2}). This value is lower than what is obtained at controlled sites. This is an indication of sub-optimal conditions for crop productivity which leads to low obtained yields.

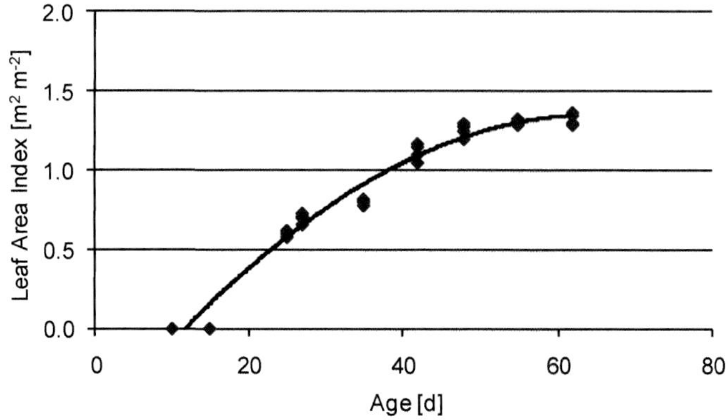

Figure 4.16 Leaf area index for the Kito maize variety as measured in Makanya catchment in 2006.

4.6.7 Obtained yields

The average maize grain yields from the four seasons and for each site and treatment are shown in Table 4.1.

Table 4.1 Grain yields obtained at each site and for each treatment over four seasons 2006-2007.

Treatment		Obtained maize grain yields (t ha^{-1})			
		Site 1	Site 2	Site 3	Site 4
Treatment 1	Range	0 - 2.24	0 – 1.53	0.88 – 4.5	0 – 3.52
	Mean	0.75	0.75	1.81	2.52
Treatment 2	Range	0 – 1.06	0 – 2.32	1.2 – 4.8	0 – 4.55
	Mean	0.51	1.36	2.46	2.82
Treatment 3	Range	0 – 3.02	0 – 1.8	1.55 – 3.8	0 – 3.17
	Mean	1.08	1.05	1.73	2.08
Treatment 4	Range	0 – 1.61	0 – 2.56	0 .05 – 3	0 – 2.81
	mean	0.58	1.31	1.27	1.80
Control	Range	0 – 0.55	0 – 1.82	0 – 1.4	0 – 3.24
	Mean	0.23	0.70	1.60	0.93

The productivity data ranges shown in Table 4.1 help to illustrate the large variability in yield obtained at each site and with treatment. Site 1 and Site 2 which did not divert much water recorded lower yields than Site 3 and Site 4 which had more efficient diversions and practiced conservation agriculture more comprehensively. Site 3, which has the largest potential for diversions, harvested at least something throughout the research period in the sections with SIs. At Site 3, only the control plot recorded zero yields in extremely poor seasons. This shows that the diversions function to provide food security.

4.7 DISCUSSION AND CONCLUSIONS

This chapter has detailed how the research sites have been selected, the challenges which are common at the research sites, and has gone further to explain the experimental design and the observations made.

The results confirm the challenges faced by rainfed agricultural farmers where rainfall is generally insufficient to support preferred food crops. However, despite the threat of obtaining sub-optimal yields every season due to water scarcity, the farmers take the same risk of growing the crops which have higher water demands than the average seasonal rainfall. A shift to more water efficient crops is difficult to achieve as the staple food is maize and rice. A normal household would aim to be self sufficient in providing staple food requirements. However, rice cannot be easily grown in such dry environments hence the focus on maize production at almost every household in the study area. A typical maize crop requires about 500 mm/season to mature (Critchley and Siegert, 1991) yet the gross seasonal rainfall at each site never reached this figure during the entire study period. This research has attempted to improve the obtained yields under these constraints.

Between 2005 and 2007, eight rainfall seasons were observed. Out of the eight, only two seasons surpassed cumulative rainfall of 200 mm at at least one site. Besides low seasonal totals, the rainfall has also been shown to be highly variable with frequent dry spells of more than 10 days between rain days. Enfors and Gordon (2007) confirm these dry spells which are largely caused by declining seasonal rainfall trends in the

Masika seasons. While daily rainfall is characterised by high intensity events of low duration, the totals are usually below 20 mm d^{-1} with a few notable exceptions of more than 80 mm d^{-1} in 2005 and 2006. In 2007 not a single day experienced rainfall exceeding 20 mm d^{-1}.

The rainfall received is converted to runoff which only lasts for short durations of up to one hour. Thereafter, particularly in small streams, the river bed dries up again. This highlights the need to harvest this water when it becomes available and concentrate the water towards productive purposes before this water is "lost" out of the farmers systems.

Lysimeter observations show that daily evaporation is about 2 mm d^{-1}. There is good agreement between manually and automatically observed data. This gives confidence to the longer data series by manual observations. The soil evaporation rates of about 2 mm d^{-1} seem on the high side. This could be a result of extra heat absorbed by the drum and that the soil in the drum was always kept moist by adding extra water when it became dry. However, the major reason is probably that semi-arid conditions always experience very high potential evaporation rates. Mul (2009) calculated potential evaporation rates of close to 10 mm d^{-1} in the study area.

An analysis of soil moisture variations as a result of the introduced techniques shows that the soil moisture levels are highest around the soil bunds while the middle section between two trenches shows low moisture levels which are comparable to the control section. This suggests that this middle section does not benefit from the ponding effect or from subsurface lateral flow especially at Site 3. At Site 4 the difference is less pronounced possibly due to the shallow soil depth and steeper slope where sub-surface flow would be more lateral. It is possible that reducing the spacing between the trenches would improve the distribution of soil moisture within the cultivated strip but this comes at a cost of the loss of land for planting the main crop (maize).

Average grain yield increases of 12.5% when extra water only is made available and which increases to 76% when manure, cover cropping and ripping techniques are applied show that a cocktail of efforts is required to transform the existing situation of low yields.

Chapter 5

WATER PARTITIONING ANALYSIS USING MODELLING TECHNIQUES

5.1 INTRODUCTION

This thesis has demonstrated, so far, that smallholder rainfed subsistence farmers are faced with the challenge of water scarcity on degraded fields. Simple and affordable techniques have been tested for their effectiveness in improving obtained yields. While on-site moisture observations have demonstrated that the tested techniques help to improve moisture availability within the root zone and, ultimately, contribute towards increased grain yields, it is also important to better understand soil moisture dynamics and the (re)partitioning of available water to better explain these incremental yields. Simple logic relates an increase in yield with an increase in green water fluxes which promote transpiration and, hence, biomass production. This is particularly true where water is the limiting factor to crop productivity but, as a matter of fact, nutrient balances and proper farm management practices play an equally important role in raising yield levels.

Modelling techniques help to better explain these soil moisture dynamics at more refined time steps. The observations made on-site provide input into the models where the modelling outputs may include simulated soil moisture and partitioning of the available water. A combination of on-site measurements and modelling approaches helps to better understand the soil moisture dynamics involving (re)partitioning of water at field scale and the associated changes in yield levels in response to changes in moisture availability.

This chapter presents the application of spreadsheet modelling techniques to understand partitioning processes associated with the techniques that have been tested in this research. The outcome from the spreadsheet modelling approach is verified with observed measurements. The numerical soil and groundwater flow model HYDRUS2D model and geophysical techniques are used to better understand sub-surface flow dynamics.

In this chapter modelling techniques that have been applied at Site 3 and Site 4, where the tested techniques were applied according to the specifications of the research, are discussed. The soils at both sites are sandy loam with depths of 1.5 - 2 m at Site 3 compared to depths of up to 1.2 m at Site 4. The shallow soils at Site 4 are largely attributed to the fact that Site 4 is on a hillslope consisting of granitic bedrock. In addition to deeper soils, Site 3 has a gentler slope of about 7% compared to Site 4 which slopes at about 12%.

5.2 WATER BALANCE MODELLING[4]

Soil moisture storage has been modelled using a spreadsheet-based water balance model (based on Savenije, 1997). The model is defined by breaking down Equation 4.1 (Chapter 4) into the following equation:

[4] Based on the following paper: Makurira, H., Savenije, H.H.G. and Uhlenbrook, S., 2009a. Modelling field scale water partitioning using on-site observations in sub-Saharan rainfed agriculture. Hydrol. Earth Syst. Sci. Discuss., 6: 5537-5563.

$$\frac{dS_s}{dt} + \frac{dS_u}{dt} + \frac{dS_g}{dt} = P - E_T - E_I - E_s - Q_g - Q_s$$ **Equation 5.1**

where (all terms in mm d^{-1}),

P is the precipitation received in the system,

E_T is the transpiration,

E_I is the evaporation from interception i.e. from canopy cover and soil surface,

E_s is the evaporation from the soil,

Q_s is the net surface runoff,

Q_g is the groundwater runoff,

$\frac{dS_s}{dt}$ is the rate of change of surface water storage,

$\frac{dS_u}{dt}$ is the rate of change of water storage in the root zone, and

$\frac{dS_g}{dt}$ is the rate of change of groundwater storage.

The components in the above equation can be determined from direct observations or, alternatively, from empirical relationships.

During the field trials, P and Q_s were measured on site as explained in Chapter 4. At the daily time scale used, $\frac{dS_s}{dt}$ is considered to be negligible compared to the other fluxes. The transpiration and soil evaporation (E_T and E_s,) are modelled as a function of the soil moisture S_s. Groundwater storage and flow ($\frac{dS_g}{dt}$ and Q_g) are fed by groundwater recharge which occurs when the soil moisture within the root zone exceeds field capacity. This study concentrates on the root zone hence these deeper groundwater processes can be ignored. The interception, E_I, is determined on the basis of the daily rainfall following the method by De Groen and Savenije (2006). These methods are described below. As a result, the soil moisture storage in the unsaturated zone, S_u, remains the only unknown in the equation. The calculated soil moisture storage is subsequently compared with the observed soil moisture variations.

Estimation of model inputs

The modelling approach for evaporation is based on the FAO-56 dual crop coefficient method which separates evaporation and transpiration processes. Figure 5.1 shows a flow chart illustrating the adapted method.

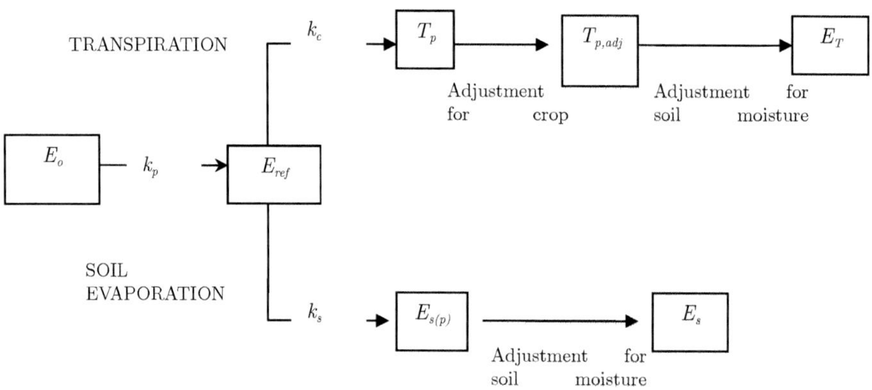

Figure 5.1 Flow chart for determining evaporation and transpiration.

Input parameters have been estimated according to the processes explained below (Allen et al., 1998; Allen et al., 2005; Savenije, 1997; Temesgen et al., 2007).

<u>Precipitation (P)</u> has been measured daily using rain gauges installed on site.

<u>Surface runoff contribution (Q_s)</u>

Runoff has been directed onto the experimental sites through one inlet point and has been allowed out through one exit point. Tipping bucket loggers have been installed at these points to measure surface flow into and out of the study plot. The difference between inflow and outflow is the net surface flow contribution, Q_s. Hence, the runoff contribution has been measured continuously.

<u>Interception (E_I)</u>

Interception is calculated as

$$E_I = \min(P,D)$$ **Equation 5.2**

where (all in mm d^{-1})

E_I evaporation from interception,

P rainfall

D interception threshold determined by calibration (ranges between 2-5 mm d^{-1})

Open water evaporation, (E_o)

Daily open water evaporation, E_o, has been measured using a Class A pan located at a nearby meteorological station. The evaporation has been determined from the measured volume required to top up the water level at a set time every day.

Reference evaporation (E_{ref})

The reference evaporation, E_{ref}, is estimated from the FAO recommended methods of estimating the total evaporation (soil evaporation, interception and transpiration) (Allen et al., 1998) i.e.

$$E_{ref} = E_o k_p$$ **Equation 5.3**

where

E_{ref} reference evaporation (mm d^{-1})

E_o open water evaporation (mm d^{-1})

k_p pan coefficient (-), ranges between 0.6-0.8 for the conditions in study area (according to FAO-56)

Transpiration

Potential transpiration, T_p

A growing crop under optimum conditions transpires at the potential transpiration rate, T_p. This potential transpiration is related to the reference transpiration by a crop transpiration factor, k_c, which is a function of the crop type and its development stage.

The potential transpiration for any crop is therefore calculated as

$$T_p = E_{ref} k_c$$ **Equation 5.4**

where

T_p potential transpiration (mm d^{-1})

k_c crop factor (-) and ranges between 0.15-1.15 for maize crop (according to FAO-56)

Equation 5.4 applies to a crop growing under ideal conditions. The natural environment necessitates a further adjustment of k_c to suit local conditions (Allen, 2000). When there is no moisture stress transpiration is assumed to be related to the leaf area index, I_{LA}, (Temesgen et al., 2007). The modified potential transpiration is hence calculated as

$$T_{p,adj} = \max\left(\left(E_{ref} k_c - E_I\right)0\right)\min\left(1, I_{LA}\right)$$ **Equation 5.5**

where

$T_{p,adj}$ adjusted potential transpiration (mm d^{-1})

I_{LA} leaf area index (m^2 m^{-2})

Actual transpiration from a crop, E_T

T_p described in Equation 5.5 assumes unlimited water availability within the root zone. In practice, however, soil moisture varies within the available water content (AWC) range described as the difference between the field capacity (S_{fc}) and the permanent wilting point (S_{wp}). Potential transpiration occurs between saturated moisture conditions until the moisture content drops to a fraction p (taken as 0.6) of the available soil moisture when stress conditions start to occur. Transpiration stops when the soil moisture level drops to the permanent wilting point. Within the moisture stress range (1-p) (S_{fc}-S_{wp}) transpiration is reduced according to proportions dictated by the gradient k which is defined as

$$k = \frac{1}{(1-p)(S_{fc} - S_{wp})}$$ **Equation 5.6**

where

k moisture stress gradient (mm^{-1})

S_{fc} soil moisture at field capacity (mm)

S_{wp} soil moisture at wilting point (mm)

p fraction of no moisture stress (-)

The moisture stress factor limiting transpiration can therefore be expressed as

$$f_{mt} = k \min((S_u - S_{wp}), 1) \qquad \textbf{Equation 5.7}$$

where

f_{mt} moisture stress factor (-)

S_u soil moisture within the root zone (mm)

The actual transpiration, E_T (mm d^{-1}), is given by the relationship

$$E_T = T_{p,adj} f_{mt} \qquad \textbf{Equation 5.8}$$

Soil evaporation, E_s

The energy available at the soil surface is shared between transpiration and direct soil evaporation (Allen, 2000). Where water is in abundance, climatic influences play a less significant role towards transpiration rates (Novák et al., 2005). As canopy cover increases, more energy is used for transpiration at the expense of direct soil evaporation. Similar to transpiration, soil evaporation only occurs at the potential rate under ideal conditions including sufficient soil moisture. Whereas transpiration occurs at reduced rates up to the wilting point, the cut-off level for soil evaporation occurs before the wilting point due to capillary forces of the soil matrix.

The soil moisture stress factor can be described by an exponential function involving S_u and the maximum water available within the root zone (S_{max}) with a reduction scale b (mm):

$$f_{ms} = \min\left(\exp\left(\frac{S_u - S_{max}}{b}\right), 1\right) \qquad \textbf{Equation 5.9}$$

where

f_{ms} moisture stress reduction factor (-)

S_{max} maximum soil moisture in the root zone (mm)

b reduction scale (mm)

The evaporation from the soil is given by

$$E_s = \max(1 - I_{LA}, 0)\max(k_s E_{ref} - E_I, 0)f_{ms}$$ **Equation 5.10**

where

E_s soil evaporation (mm d^{-1})

k_s soil evaporation factor (equivalent to crop factor in cropped areas) (-)

In Equation 5.10, interception is subtracted from reference evaporation because both are evaporation processes with evaporation from interception occurring immediately after a rainfall event as canopy interception or evaporation from the soil surface.

Infiltration, F, and deeep percolation, R

At daily time steps, where $\frac{dS_s}{dt}$ is considered negligible, the infiltration into the soil F (mm d^{-1}) is calculated as

$$F = P + Q_s - E_I$$ **Equation 5.11**

The soil moisture balance at any given time-step t is hence calculated as

$$\frac{dS_s}{dt} = F - E_T - E_s - R$$ **Equation 5.12**

Where, deep percolation, R (mm d^{-1}) is calculated as a flow over threshold process which only occurs when the field capacity is exceeded.

$$R = \max\left[\frac{S_u - S_{fc}}{k_R}, 0\right]$$ **Equation 5.13**

and k_R (d) is the maximum number of days during which field capacity can be exceeded after high infiltration events.

5.2.1 Results

Soil moisture observations

Results from soil moisture observations have been analysed in Section 4.6.4 (Chapter 4). In general, Tube D records the highest soil moisture levels which is mainly a function of ponding effects around the location of Tube D. Tubes A and C record the least soil moisture levels while Tube B records moisture levels which are in between.

Soil moisture modelling

The spreadsheet based water balance model has been constructed as conceptualised above. The model simulates soil moisture on a daily time step. Each simulation calculates soil moisture for the control and the experimental site (Treatment 1) with the difference being that the experimental site allows extra water from diversions while the control is strictly rainfed. The output is plotted in Figure 5.2 where the solid lines indicate the simulated soil moisture.

The modelling results visually show a good agreement with the observed soil moisture for both the control and the portion benefitting from diverted water. The Nash and Sutcliffe coefficiency of efficiency has been determined to test the level of confidence between the observed against the modelled results. The Nash and Sutcliffe coefficient (E_f) ranges between 1 and $-\infty$. A value of 1 indicates a perfect relationship while a value of 0 indicates that the modelled results are only as good as the mean of the observations. The modelled results for diverted flow correspond well with the downslope tube (Tube D) with E_f values of 0.94 and 0.8 at Site 3 and Site 4, respectively. The simulation with no diversion corresponds well with the control (Tube A) with E_f values of 0.63 and 0.53 (which improved to 0.84 after model refinement) at Site 3 and Site 4, respectively. The difference between the control and experimental lines indicates the benefit of the diversion technique. The biggest difference occurs at the beginning of the season and is lowest when enough rainfall is realised and field capacity is attained. At both sites the simulated flow with diversions shows that the soil reaches field capacity much earlier than the control section at the onset of the rainy season. This is important since it allows the growing season to start earlier.

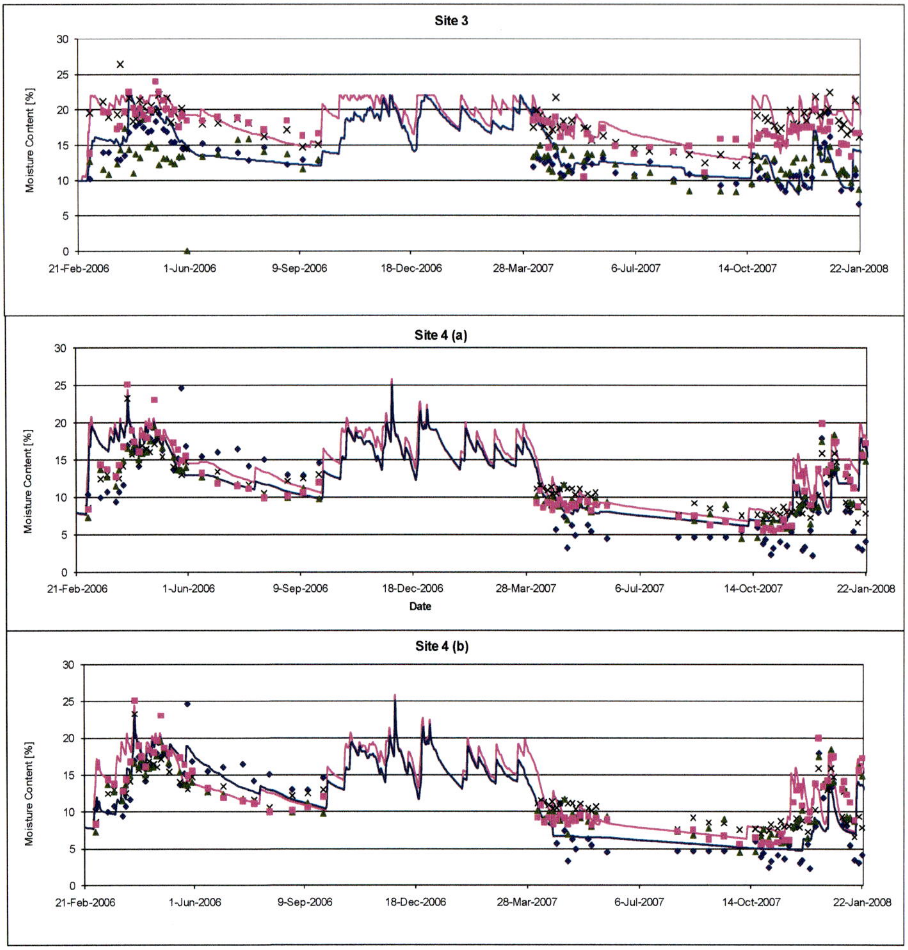

Legend

simulated — simulated control ♦ Tube A x Tube B ▲ Tube C ■ Tube D

Figure 5.2 Model results at different sites compared with observed soil moisture values for Site 3 and Site 4. The graph at Site 4(b) presents an improved simulation than the graph at Site 4(a).

Improving model performance

Figure 5.2, Site 3 shows better agreement between observed and simulated soil moisture levels than Site 4. The simulation (a) of Site 4 shows disagreement especially at the onset of the experiment where simulated values are much higher than the observed. This is explained by the fact that the experimental plot may not have been prepared well enough with the top soil still very dry hence less infiltration actually occurred. The model was improved by lowering the field capacity during the first few days of experimentation thereby restricting infiltration after rainfall events. Thereafter, and for the rest of the season, the control plot recorded higher moisture levels as it benefitted from lateral flows from upslope. Similarly, during the dry season prior to November 2007, a few rainfall events were observed which were translated into infiltration in the simulation. Consequently, simulated results were much higher than the observed. However, since the previous rainfall season had recorded extended dry spells towards the end, the soils were in fact much drier hence, again, the rainfall could not practically be translated into infiltration. The model was corrected by draining this infiltrated water during the dry season.

The improved model output is shown as Site 4 (b) in Figure 5.2 and shows a much better correspondence between modelled and observed values.

Confirmation of model improvement

Figure 5.3 shows a comparison of the modelled and observed results with the graph forced to pass through the origin. All trend lines have slopes close to 1 which is satisfactory. Site 4 (b), which shows the trends after the model improvement described above, reflects an improvement in R^2 value which reflects more refined simulations. This is also confirmed by the Root Mean Square Error (RMSE) calculation of 3.3 and 1.9 at Site 3 for Tube A and Tube D, respectively. At Site 4, the RMSE decreased from 4.2 to 2.6 for Tube A after the model improvement while it increased from 2.3 to 2.4 for Tube D. This also shows that the model improvement was most efficient in the control plot.

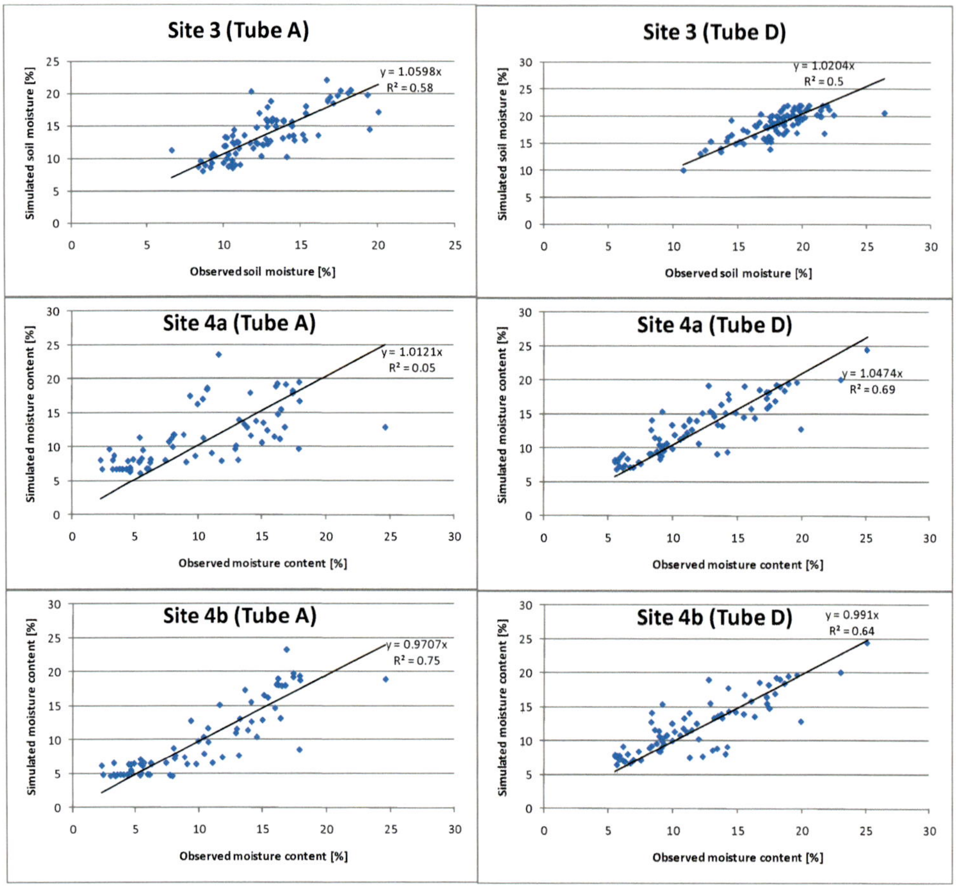

Figure 5.3 Comparison of observed and modelled moisture in Tube A (control) and Tube D.

Sensitivity analysis

A sensitivity analysis has been conducted to check if the assumed values for some parameters used in the model would have any significant influence on model output. Assumptions were made for p (soil moisture depletion factor), k (residence time of water within soil profile above field capacity), D (Interception threshold), k_c (crop coefficients) and k_p (pan factor).

k_c and k_p values are obtained from standard guidelines. Also p is not likely to vary much away from the generally recommended value of 0.6. These parameters result in minimum sensitivity within reasonable ranges as offered in standard guidelines. k_R

(residence time above field capacity) does not affect the daily water balance as, in the model, this retention time does not exceed 1.5 days. This only leaves the interception threshold as an important parameter to be tested in a sensitivity analysis. D ranges between 2 - 5 mm d^{-1} (after De Groen and Savenije, 2006).

The sensitivity analysis shows that D is not a sensitive parameter for the calculation of transpiration values. Figure 5.4 shows graphs of total seasonal transpiration obtained for interception values ranging between 1 - 5 mm d^{-1}, and a comparison of the values with the 3 mm d^{-1} interception threshold used in the model. Seasonal transpiration for D values of 1, 3 and 5 mm d^{-1} (D1, D3 and D5, respectively) are plotted against the value used in the model, D3 (i.e. a change in D (dD/D) of 67%). This change in D would result in change in transpiration of dT/T.

The deviation from the D3 graph is less than 20% in all cases (See Figure 5.4). The sensitivity of the relationship can be expressed in as an elasticity relationship [(dT/T)/(dD/D)] where an elasticity of 1 reflects a highly sensitive relationship. In most cases, an inelastic relationship of less than 0.3 is obtained and which confirms that D is not a very sensitive parameter. While the interception threshold does not prove to be a sensitive parameter, interception is still important in water balance analysis as the available water for other processes, e.g. transpiration, is depended on the balance available after interception.

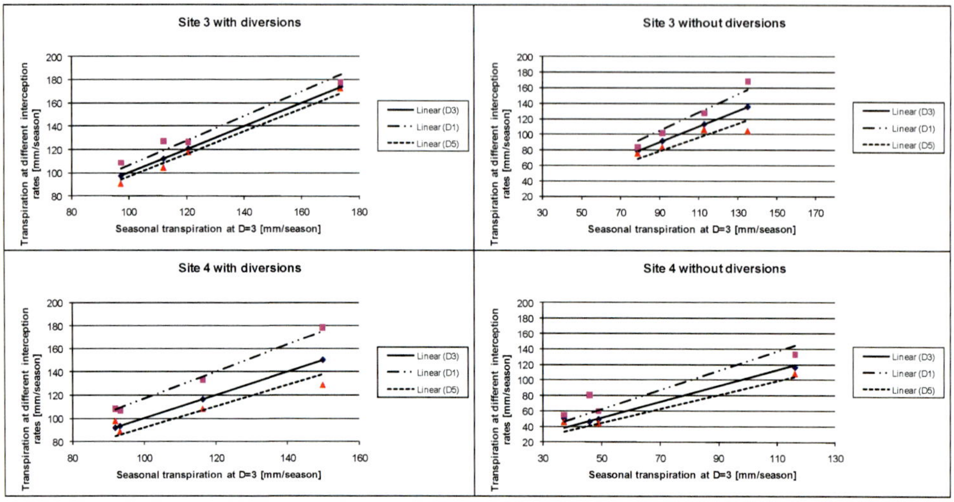

Figure 5.4 Comparison of seasonal transpiration values for different interception thresholds.

5.2.2 Water balances

The total amount of water received at field scale is due to the rainfall (P) and runoff diversion (Q_s). This water is partitioned into transpiration (E_T), interception (E_I), soil evaporation (E_s), deep percolation (R) and soil moisture storage variation ($\mathrm{d}S_u/\mathrm{d}t$).

Table 5.1 shows the average daily water partitioning for the combined cropping seasons. At Site 3, the volume of water diverted surpasses the rainfall received while at Site 4, the runoff contribution is about 30% of rainfall received. Deep percolation accounts for the largest proportion of the partitioned water.

For the combined seasons, at Site 3, transpiration increased from an average of 1 mm d^{-1} to 1.4 mm d^{-1} (range 0.86 - 1.93 mm d^{-1}) as a result of the improved agricultural techniques. At Site 4, the average transpiration increased from 0.7 mm d^{-1} to 1.1 mm d^{-1} (range 0.37 - 1.26 mm d^{-1}). Despite an increase in transpiration values, deep percolation accounts for almost 50% of the diverted water at both sites.

Table 5.1 Water partitioning "with" and "without" SIs (mm d^{-1}).

	P	Q_s	Total inflows		E_T	E_I	E_s	R_g	$\frac{dS_u}{dt}$	Total outflows
Site 3										
with	2.1	2.7	4.8		1.4	0.7	0.2	2.3	0.2	4.8
without	2.1	0	2.1		1.0	0.7	0.1	0.2	0.1	2.1
Site 4										
with	2.8	1.0	3.8		1.1	0.7	0.2	1.6	0.2	3.8
without	2.8	0	2.8		0.7	0.7	0.2	1.0	0.2	2.8

5.2.3 Analysis and discussion of results

The model results show the positive effects of the tested diversions. The diversions and temporary in-field storage structures shorten the time it takes to attain sufficient moisture levels for germination, thus effectively lengthening the growing season. This means that crops grown under the adjusted farming conditions have a longer growing season, less chance of suffering from moisture stress during dry spells and, hence, stand a higher yields compared to traditional practice. Grain yield increases of more than threefold have been recorded under these improved farming systems (Makurira et al., submitted). At Site 3, the difference in moisture availability between control conditions and tested techniques is much higher than at the other site due to the fact that the diversion potential is much higher at Site 3 and, also, the gentler slope at Site 3 promotes more water retention compared to Site 4.

The moisture gap between the control and the diversion site in the dry season (Figure 5.2) suggests that residual moisture is higher under the new technique, thus allowing for the cultivation of alternative crops in the dry season, particularly in the trench. These dry season crops have proved a success and provide additional food in the dry season.

The water balance analysis shows that the effect of the diversion is also a function of slope and soil depth. Steeper slopes result in more runoff from the system through lateral flow. Diversions result in more water available for productive purposes but the proportion of water attributed to deep percolation also demonstrates the fact that in these situations, where rainfall and runoff events are of short duration, the generated flow cannot all be absorbed in the root zone within such short periods. In this case, the trenches obviously do not offer sufficient storage to regulate the release of water into the root zone when required. Rockström (2001) also showed that non-productive purposes (evaporation and deep percolation) can easily account for more than 50% of the available water under comparable conditions.

5.2.4 Conclusions

The effect of the innovations introduced in the farm plot (*fanya juus*, runoff diversion and conservation tillage) have been modelled successfully and it has been shown that the advantage of the tested innovations is that it allows for the growing season to start early and contributes towards dry spell mitigation by raising soil moisture levels. It has also been demonstrated that the biggest impact of the *fanya juu* terracing in combination with diversions is through ponded water around the soil bunds. Where soils are not deep and with steeper slopes, the findings show that the water in the trenches is transferred to sub-surface lateral flow which is not available to the crop. Since the trenches and bunds were constructed according to soil conservation guidelines, this then may suggest that, in steeper slopes and for the *fanya juu* structures to be more effective, the spacing between the bunds and the trenches may have to be less than that recommended for soil conservation purposes.

The tested techniques allow for higher moisture levels even in the dry season which is an advantage especially for the longer season alternative crops (such as bananas, pawpaws, fodder and cassava) which are grown in the trench and at the bunds. However, the general decline in soil moisture levels throughout the dry season as a result of soil evaporation implies that valuable soil moisture is lost through soil evaporation during the dry season. If conserved, the moisture level at the close of the growing season could provide a better starting point at the beginning of the following season. A way of minimising dry season evaporation can significantly benefit the performance of the subsequent season. Future research should focus on investigating different land management techniques (e.g. different ploughing techniques, or reducing bare soil evaporation by introducing a minor crop for canopy or by covering the soil with mulch, or by breaking the capillary rise (Wallace et al., 1999) to reduce moisture losses through soil evaporation during the dry seasons.

The high proportion of water entering the deep percolation zone suggests that there is still inefficient utilisation of harvested water at the investigated field plots. In hydrological terms this is not a loss as this water would most likely be used further downstream. However, at local scales, this demonstrates that the in-field temporary storages created cannot cope with the generated volumes of water and do not allow for regulated release of water into the root zone when required. This suggests that the tested techniques can perform even better when used in combination with storage systems and soil moisture retention techniques for more effective dry spell management. Future investigation should focus on the tested techniques in combination with micro dams and/or storage tanks for rainwater harvesting.

5.3 APPLICATION OF THE HYDRUS2D MODEL TO INTERPRET SUB-SURFACE FLOW DYNAMICS

5.3.1 Background to the HYDRUS2D model

The HYDRUS2D model is applied to simulate unsaturated sub-surface flow based on the Richards equation (Simunek et al., 2006; Verbist et al., 2009). Soil moisture flow depends on the soil characteristics, slope, the rate of injection of water into the soil structure and the prevailing boundary conditions. The advantage of HYDRUS2D is that it allows for the creation of a fine grid mesh through which water movements are monitored together with water balances within the fine mesh. It also works for different geometries as specified according to field conditions. In this research, the model has been run on daily time steps.

5.3.2 Model setup and inputs

Simulations have been performed for the 2006 long rainfall season in Makanya catchment. The modeling process concentrated on the soil moisture dynamics within the profile over the season. The geometry of the profile has been drawn to scale showing at least two cultivation strips between the *fanya juu* constructions. The following scenarios were simulated: (a) natural conditions (control), (b) *fanya juu* constructions without storm diversions at the existing spacing of about 10 m, (c) *fanya juu* constructions with storm diversions at the existing spacing, (d) *fanya juus* with reduced spacing of 3 m, and, (e) *fanya juu* at spacing of 20 m.

At both sites a soil depth of 2 m has been used for sandy loam profiles but with different boundary conditions. A seepage face lower boundary condition has been applied for Site 3 while, at Site 4, a no flux boundary condition has been applied at the bottom surface. A variable head boundary condition was applied to the trenches to allow them to simulate the diverted water.

Since the primary focus of the research has not been on detailed soil physics, but rather, on water partitioning, it was considered out of scope to conduct detailed soil tests on soil hydraulic properties. The default settings for soil hydraulic properties for the given soil texture which are in-built in the model have been adopted.

Daily observed rainfall data has been used. Because the model cannot handle large water influxes, the daily rainfall was distributed over quarterly periods within the day. Water harvested from storm flow diversions has been incorporated into the storm through assigning a variable head boundary condition to the trenches.

The performance of the model has been compared against a spreadsheet water balance model (SModel) which has been run at the same locations (Makurira et al., 2009a).

5.3.3 Results

The results from the HYDRUS simulations show moisture progression within a soil matrix at different times. Figure 5.5 shows the typical distribution of soil moisture for the various scenarios on a random day (day 51) of the growing season.

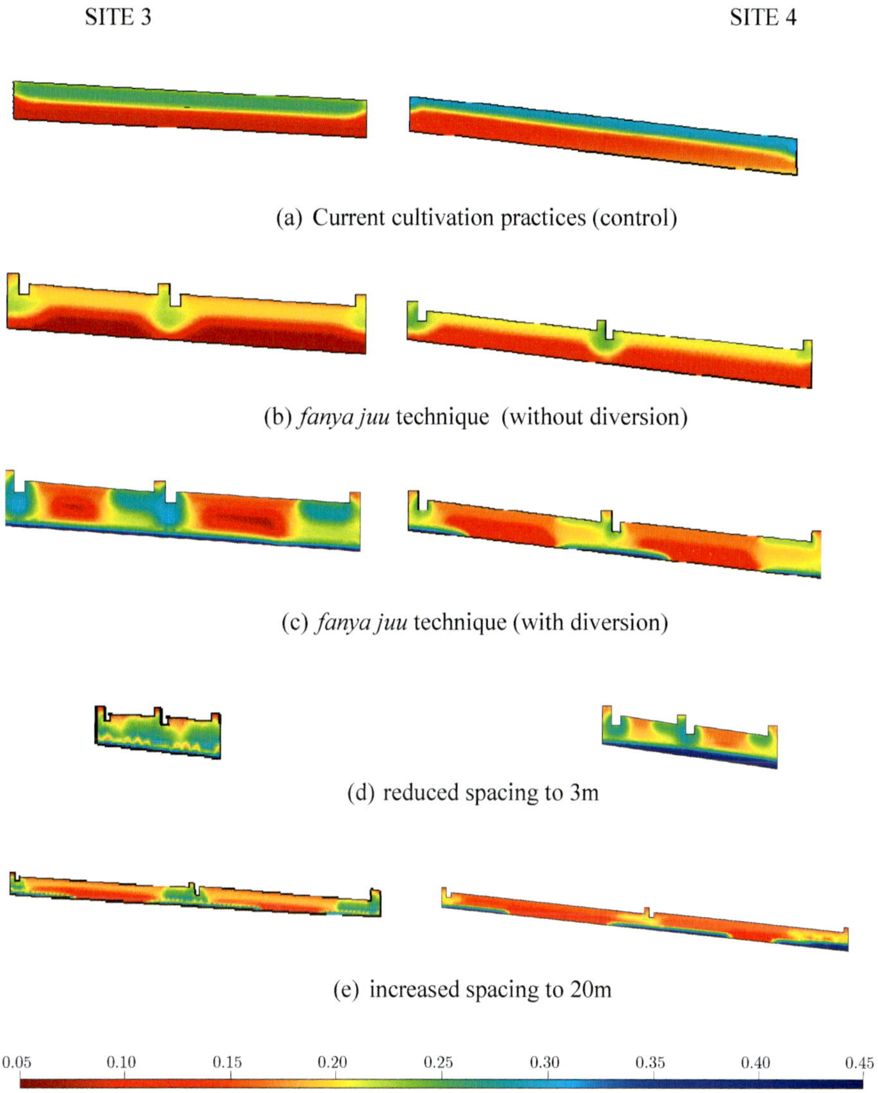

Figure 5.5 Pattern of soil moisture variation at Sites 1 and 2 with different field layouts.

Although the rainfall received at each of the two sites is different, similar patterns appear within the soil matrix. The pictures show that, under the traditional tillage systems (control), the soil moisture pattern is homogeneous in the lateral direction which implies that the dominant sub-surface fluxes are vertical. However, Site 4 shows some evidence of lateral drainage which are influenced by the slope. When *fanya juu* trenches are introduced without diversion, there is no evident improvement in soil moisture levels as the little water available is concentrated into the trenches thus leaving very little water to infiltrate in the cultivated section. When water is diverted into the field, both sites demonstrate a significant increase in soil moisture which is influenced by the trenches and the ponding from the bunds. However, the pieces of land between the trench and the bund exhibit very little moisture increase. This shows that ordinary infiltration results in much less water availability in the root zone. Infiltration potential is increased by the trenches and the ponds.

Simulations to investigate the effect of spacing show that moisture is more equally spread, if the spacing between the trenches is reduced to 3 m and that it is apparently localised only to the trench and bund when this spacing is increased to 20 m. This suggests that, at short time steps, lateral soil fluxes are insignificant for the given slope and soil material.

Observation points within the soil profile monitor soil moisture variation. Figure 5.6 shows the variations of soil moisture across the cultivated strip at 50 cm below the surface as computed by the HYDRUS2D model. Point B corresponds to the location of Tube B and is closest to the trench; point C relates to the middle tube while point D is nearest to the bund. The line "avgBCD" reflects the average soil moisture recorded by the three tubes B, C and D while the line "avg BD" reflects the average soil moisture recorded by tubes B and D which are believed to benefit more from the *fanya juu* constructions.

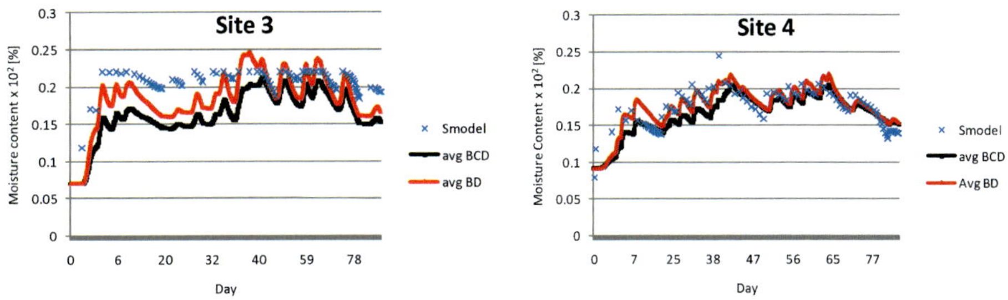

(a) (b)

Figure 5.6 Soil moisture variations with time and location as computed by HYDRUS2D and compared with the output from the hydrological spreadsheet model (SModel).

The SModel values show the soil moisture variation modelled using the spreadsheet model described in Section 5.2. The spreadsheet model has proved to perform well against observed data (Makurira et al., 2009a). The SModel curve agrees strongly with curve D at Site 4 (Walter) while at Site 3 (Iddi), although it is again related to curve D, the peaks in the HYDRUS model seem to be more exaggerated.

Figure 5.7 shows the simulations by HYDRUS2D of soil moisture variation across the cultivated portion between the *fanya juus*. There is no difference between the moisture availability with or without diversions at location B which suggests that the water stored in the trenches does not immediately contribute towards the moisture content in the root zone of the cultivated strip. A similar pattern is observed at location C where there is minimum benefit from infiltration, subsurface lateral flow and ponding effects. For point D, however, the effects of diversion are much more visible. The distance between the control and diverted graphs shows the benefit of diverting water into the field. With the diversion, it shows that the moisture content around zone D quickly rises soon after the early rains which is good for germination and dry spell mitigation.

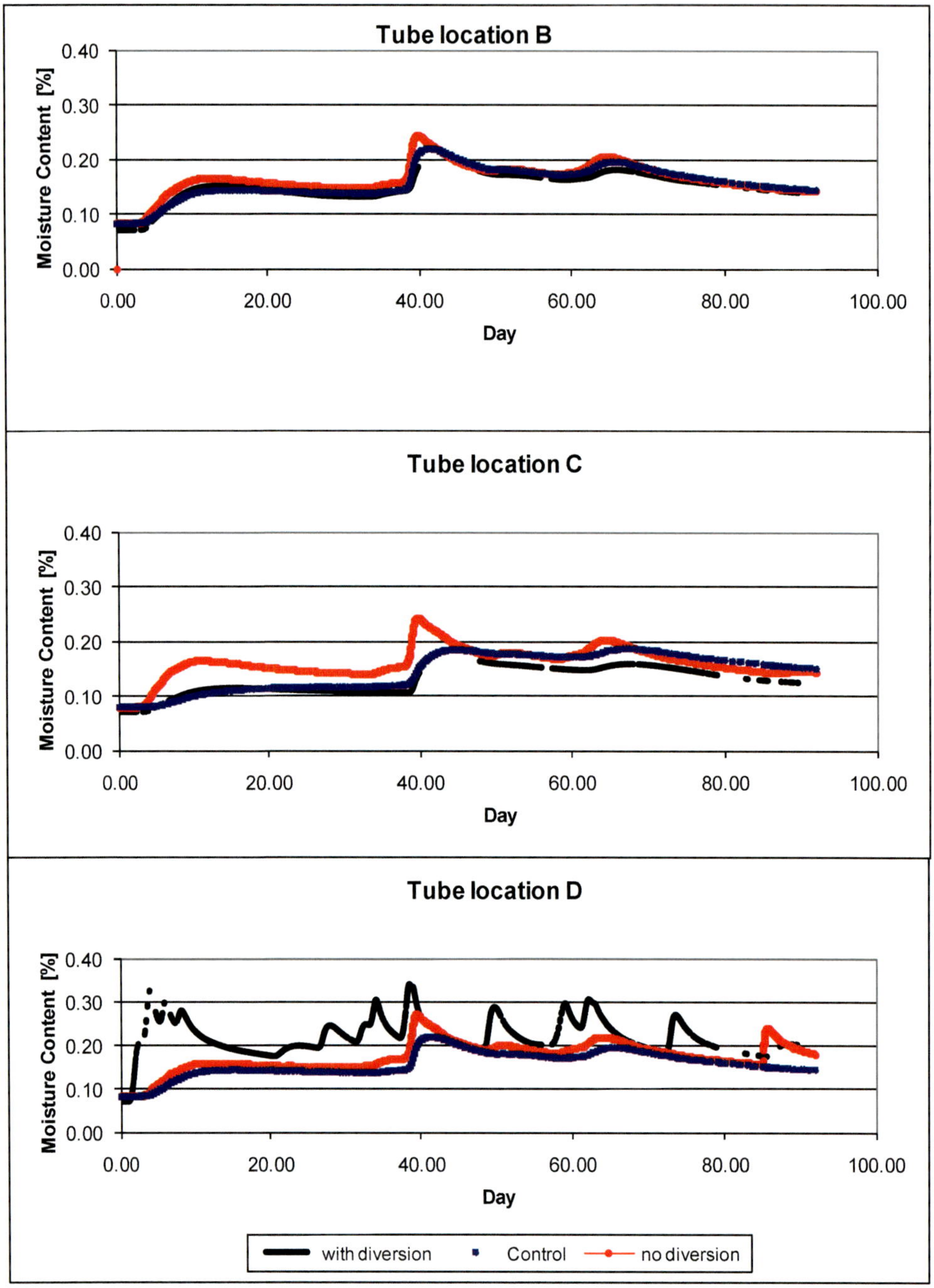

Figure 5.7 Impact of *fanya juus* and diversions at different observation points across the cultivated strip at Site 3.

5.3.4 Discussion and conclusions

The results show that current agricultural practices are not efficient in diverting water to the root zone. The *fanya juus* have demonstrated promise only when used in combination with runoff diversion. The HYDRUS2D model shows that the highest moisture concentrations occur around the bunds and trenches especially after runoff events. This water eventually drains as deep percolation or lateral sub-surface flow. The ponding effect therefore contributes most to soil moisture retention. A shorter spacing between the *fanya juu* constructions therefore shows greater potential to concentrate more water in the cultivated strip but this would be at the expense of land for farming (if most popular crops like maize are to be grown). However, from a food security perspective, this loss of land to staple food crops is compensated by growing alternative crops such as pawpaws, bananas, cassava and fodder around the trenches and bunds. These crops perform well even during the dry seasons hence provide source of food even during dry seasons. More trenches also entail more labour input. On the other hand, wider spacing between the trenches would only result in localised moisture concentrations around the bunds and trenches only.

The steep rise in moisture availability at zone D (Figure 5.7) when diversions are effected shows the efficiency of the tested techniques in attaining sufficient moisture conditions for crop growth in the early stages of the season which is important in such unpredictable climatic conditions. On the other hand, while it takes a few days to attain soil moisture levels of at least 20% with diversions and *fanya juus*, under current cultivation practices this took about 30 days to achieve. It means that, under traditional farming systems, a crop would have been deprived of about 30 days of conducive moisture conditions for growth. This ultimately impacts on final yields. This positive alteration to the water balance has been confirmed by different approaches i.e. modelling approaches (water balance and HYDRUS2D modelling), geophysical investigations and direct observations from repeated measurements.

The performance of the model is in close agreement with the spreadsheet model as earlier presented in this chapter. Minor variations have been observed which could have arisen from the fact that model default values were used for soil properties. If actual soil data had been available, the model accuracy could have improved. The HYDRUS model series contains complicated soil moisture flux series (Pachepsky et al., 2006) and this requires more time investment, especially, in soils investigations. However, a balance needs to be struck between making further model improvements and the practical constraints in measuring soil hydraulic properties (Ndiaye et al., 2007).

The HYDRUS2D model has been successfully applied to demonstrate the effectiveness of the tested techniques in the Makanya semi-arid region. The research has applied this model to demonstrate the effectiveness of the tested techniques and to better understand soil moisture dynamics as a result of the diversions and *fanya juu* constructions. For these techniques to be most efficient, the spacing between the trenches should be at minimum levels. The impacts of upscaling such innovations have not been explored in this research.

The greatest benefit from the tested techniques is derived from the ponding effect while water captured in the trenches is likely to drain vertically and recharge the groundwater. Therefore, unless if plants are grown in the trenches to tap into the high moisture zones created, the trenches would not serve much agro-hydrological purpose at local scales. However, from a wider hydrological perspective, this deep drainage recharges groundwater which becomes available for downstream uses hence is a gain to the overall ecosystem.

5.4 APPLICATION OF GEOPHYSICAL METHODS AND REPEATED SOIL MOISTURE MEASUREMENTS TO INTERPRET SUB-SURFACE FLOW DYNAMICS

5.4.1 Introduction

The accurate analysis of water flow pathways in a drainage basin is essential for the optimal protection and management of surface water and groundwater resources and the understanding of in-stream abiotic conditions (Wenninger et al., 2008). Field scale moisture dynamics are more important to smallholder farming systems where a combination of poor soils and in-field water scarcity due to highly variable rainfall seasons result in low yield for most common crops (Makurira et al., 2007b). A big challenge therefore exists to find out if there are ways of altering the present inefficient water partitioning processes in favour of productive transpiration and, hence, biomass production.

In soil and agronomic science, soil moisture dynamics are closely related to the interactions between hydro-meteorological conditions and the relevant soil characteristics. On-site observations are useful to establish prevailing soil and water conditions to support crop growth. However, it is not always feasible to conduct comprehensive on-site measurements as equipment, time and staffing constraints usually discourage these efforts. Besides, repeated measurements in cultivated fields alter soil characteristics through excavations or compaction. Non-destructive and rapid measurement options exist (e.g. Koch et al., 2008) and are sometimes preferred particularly if the objective is not to obtain absolute soil moisture values but, more generally, to quickly understand the soil moisture dynamics without disturbing in-situ conditions. The Electrical Resistivity Tomography (ERT) technique is one typical example of geophysical techniques that can be used to indirectly determine soil moisture responses to infiltration events (e.g. Asfahani, 2007; Binley et al., 2005; Uhlenbrook et al., 2005). The ERT technique involves sending electrical signals which return a measure of the resistivity of the underlying rock or soil structure up to a specified depth. When repeated over time and over the same points, the difference in resistivity can only be attributed to soil moisture changes at that point.

This section discusses the application of geophysical techniques to provide a deeper understanding of soil moisture variation in the root zone in response to rainfall input for smallholder rainfed agricultural systems. Repeated soil moisture observations are used to verify the output from the ERT methods.

5.4.2 Direct soil moisture monitoring

Time Domain Reflectory (TDR) access tubes have been installed along a transect following the slope. The first tube has been placed at the most upslope part of the field, the control, where traditional farming practices are maintained and with no benefit from additional water from diversions. At each site, four access tubes have been installed with one tube located in the control section (Tube A). Tubes B, C and

D have been placed within one cultivation strip bordered by the trenches as shown in Figure 4.6 (Chapter 4).

Downslope and along the transect, subsequent TDR tubes have been placed at the centre of each cultivated corresponding to the position of Tube C. According to the design, the diverted water ponds in the trenches first before overflowing into the downslope bunds. The water ponds at the bund first before the excess water spills into the next trench. Tube B therefore monitors the contribution of the trenches to soil moisture while Tube D monitors the effect of the ponding on soil moisture distribution. Tube C monitors the soil moisture distribution in the middle of the cultivated strip. In practice, however, excessive and unexpected diversions may break the bunds resulting in uncontrolled flooding. It is therefore important that someone monitors the field during diversion events.

Soil moisture has been monitored using probes inserted into the tubes with measurements taken at 10cm depth intervals. These measurements have been taken twice weekly in the wet season and once fortnightly in the dry season.

5.4.3 Indirect soil moisture monitoring (ERT)

In the same transect, the ERT technique has been applied as an indirect method to monitor soil moisture variation along the same transect as TDR observations. The Syscal Junior Kid Switch equipment has been used to monitor the variation in resistivity in the soil using the dipole-dipole setting based on Loke (2003) and Koch et al. (2008). In this research the upper 2 m of the soil structure have been of major interest as they cover the rooting depth of many crops. Electrodes have been placed at 2 m spacing along the measured transect for, mostly, 36 measured points (i.e. one cable length and one roll-along). Level 9 for depth of measurement has been selected. All other input assumed default values for the dipole-dipole setting that was selected.

5.4.4 Data processing

Direct soil moisture measurements

The data collected from direct soil moisture observations during the 2006 *Masika* season have been plotted to show soil moisture variation over time for each location. In the analysis, box plots have been constructed for the four tubes to show the moisture level variations in response to the functioning of the *fanya juus* at each location.

Indirect soil moisture measurements (ERT)

It is a big challenge to interpret the images correctly as the interaction between the signal emitted and the subsurface material is not always straightforward (e.g. Kock et al, 2008). The PROSYS software package has been used to process the geophysical data into measured apparent resistivities. The PROSYS program allows for the filtering of bad data points. At this stage, all data with resistivities in excess 5000 Ω.m and/or standard deviation of more than 10 Ω.m have been eliminated. The filtered files in PROSYS have then been exported to an ASCII file containing the measured apparent resistivities along the transect. These measured apparent resistivities have then been read into the RES2DINV software package (Loke, 2003). The RES2DINV checks for inconsistencies in the measured apparent resistivities where bad datum points can be further exterminated. The "refined" apparent resistivities have then been processed in RESD2DINV by inverse modelling techniques to produce a model of the resistivities within the defined soil block. The electrode spacing of 2 m used in the field translates to sounding depths of up to 9 m. Model refinement in RES2DINV allowed for the analysis of a reduced electrode spacing of 1 m.

5.4.5 Results

Direct soil moisture measurements

Figure 5.8 shows box plots of soil moisture variations along the longitudinal section during wet seasons. The boxes are in the sequence of Tubes A-D as presented in Figure 4.6 (Chapter 4). At Site 3, the median values for soil moisture are 12.5%, 17.5%, 12.5% and 19%, for Tube A, B, C and D, respectively. At Site 4 the corresponding values are 8%, 9%, 10% and 12%, respectively. This confirms that, over many seasons, Site 3 records more moisture than Site 4 and that the variance of moisture is much less at Site 3 than at Site 4. Also, at Site 3, there is a clear variation of moisture content with respect to its location between the *fanya juu* constructions. Tubes A and C appear not to respond significantly to the rainfall/runoff events while Tubes B and D respond favourably to moisture additions. However, for Site 4 (which has a steeper slope), although Tube D again records the highest moisture levels, the moisture distribution pattern appears to increase gradually with slope with larger variations in observed moisture. A test for significance ($p < 0.05$) of the differences in the mean between the control tube (Tube A) and the other tubes are statistically signifannt for all cases except for Tube C in case (a).

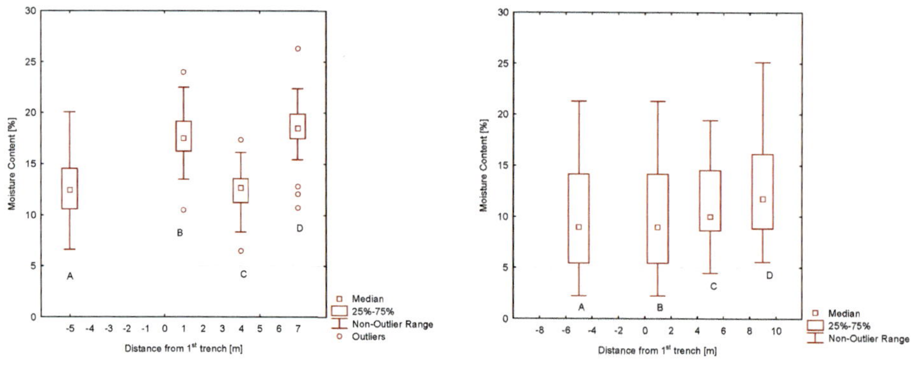

(a) (b)

Figure 5.8 Box plots of soil moisture variation between *fanya juu* constructions in all wet seasons during 2006 - 2008 at (a) Site 3 and (b) Site 4.

Figure 5.9 shows the moisture distribution during the dry seasons. At Site 3 the soil moisture distribution is similar to the wet seasons while at Site 4 there is a decrease of soil moisture distribution with slope. No statistical significance was observed between the differences in the mean of Tube A and the rest of the tubes except for Tube B and Tube D in case (a). However, it is interesting to note that, at both sites, Tube D, which shows higher moisture levels in the wet season records comparably less moisture during the dry season and the average moisture level is lower than at Tube B in both cases. This can be explained by the fact that longer season and deeper rooted alternative crops (e.g. bananas, cassava, pawpaws and sugarcane) have been planted at the bunds (near Tube D). These transpire even during the dry season thereby depleting soil moisture at faster rates than anywhere else across the field. Another possible explanation could be that the trenches receive a lot of water during diversion events which then slowly drains laterally even during the dry season. In that case, Tube B would benefit most from such moisture transitions.

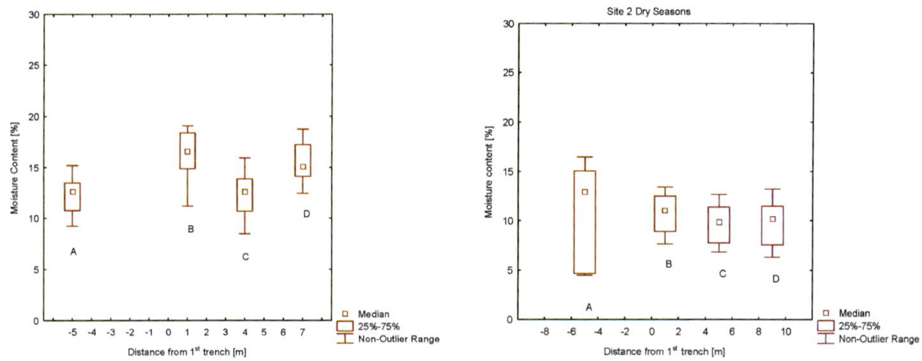

(a) (b)

Figure 5.9 Box plots of soil moisture variations between *fanya juu* constructions in all dry seasons at (a) Site 3 and (b) Site 4 during the period 2006-2008.

Figure 5.10 shows the wet season soil moisture variation at the centre of the cultivated strip in a downslope transect. At Site 3, the soil moisture distribution is uniform down the slope while, at Site 4, there is an indication of soil moisture increase with slope. There is significant difference of the means of the moisture values ($p < 0.05$) in all cases except for Tube E case (a).

A similar pattern is observed in Figure 5.11 where the dry season moisture variation at the same locations is presented. This suggests that more lateral distribution of the diverted water occurs at Site 4 then at Site 3, which is attributed mainly to the difference in slope and shallower soil depths at Site 4. No significant difference has been observed for the difference in the mean moisture values between the control snd all tubes except for Tube C in case (b).

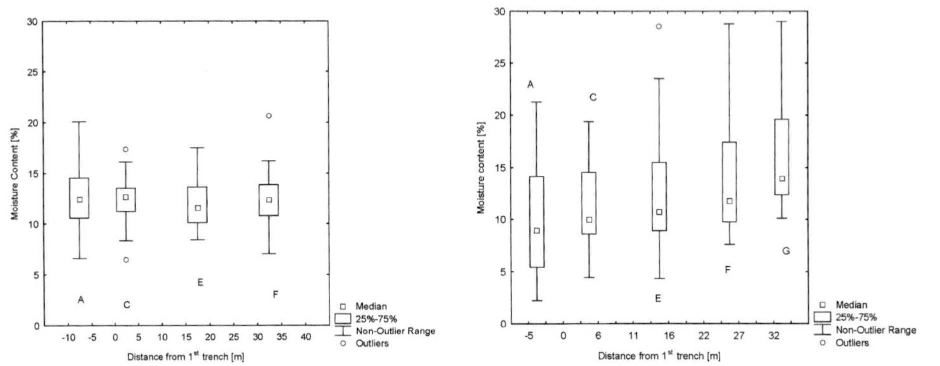

(a) (b)

Figure 5.10 Box plots of soil moisture variation in all wet seasons (2006-2008) at (a) Site 3 and (b) Site 4 for centre tubes.

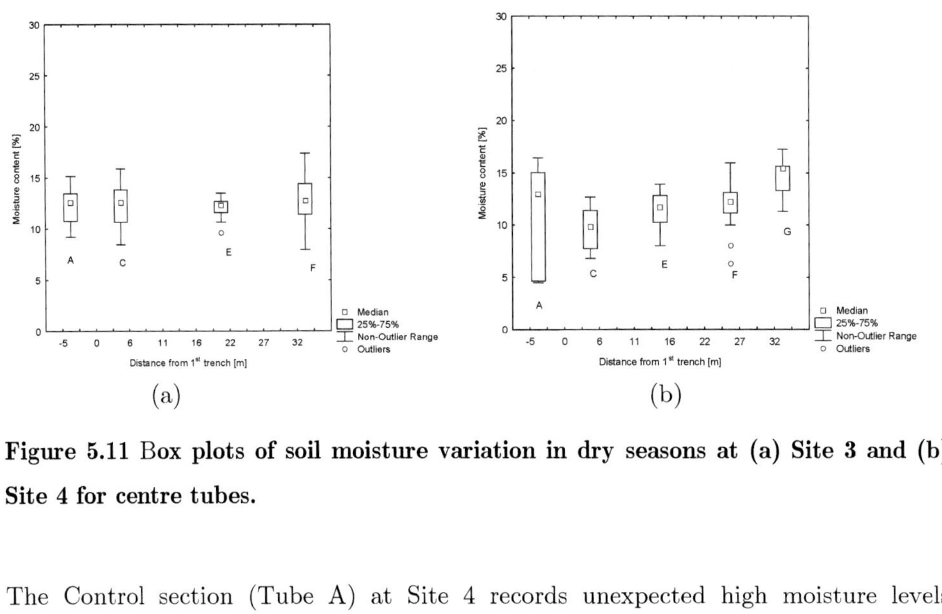

(a) (b)

Figure 5.11 Box plots of soil moisture variation in dry seasons at (a) Site 3 and (b) Site 4 for centre tubes.

The Control section (Tube A) at Site 4 records unexpected high moisture levels within the plot especially during the dry season. This suggests that the soils at the control section have higher moisture holding properties than anywhere else in the field. This moisture holding property gives an indication why, in average seasons, the yields obtained at Site 4 are higher than at the experimental sections. Possibly, this is due to the fact that the control section attains sufficient moisture conditions before other positions within the field or that the control section has different soil characteristics which include higher water holding capacities.

ERT results

After filtering out bad data points, only 48% of the total data points collected from a possible 1872 points at Site 3 remained for analysis while at Site 4, only 33% out of a possible 1224 of the data that was initially remained. This shows that the data collected was of poor quality which can be a reflection of the method used for sounding (dipole-diplole setting) and/or that the electrode spacing of 2 m created a lot of noise in the data as it had to penetrate depths of up to 9 m when the bedrock is at shallower depths of around 2 m. Only daily data which, after filtering, still retained at least 80% of the original daily data has been used in further analysis. This translates to only one day at Site 3 and two days at Site 4 of successfully observed points.

The ERT pictures presented in this section show an analysis of the apparent resistivities at the moment of measurement. The dotted vertical lines show the approximate location of the *fanya juu* constructions. The resistivity ranges observed are below 300 Ω.m. Asfahani (2007) recorded an average resistivity range of 380 Ω.m in basalt material. Figure 5.12 shows the picture obtained on 11 April 2006 at Site 3. From the picture, it is expected that the locations of the *fanya juu* constructions form wetter zones from where infiltration into the deeper soils is concentrated.

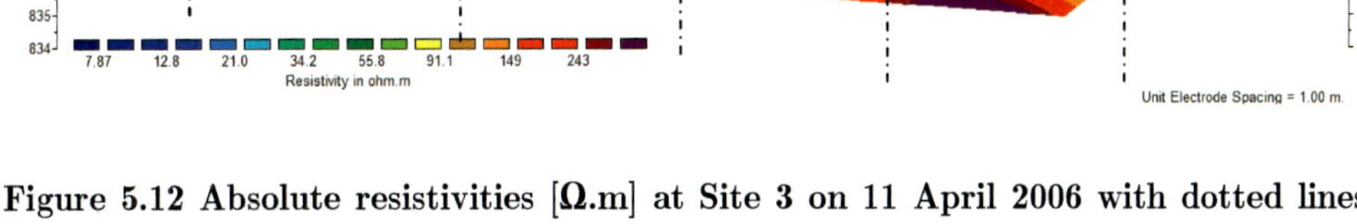

Figure 5.12 Absolute resistivities [Ω.m] at Site 3 on 11 April 2006 with dotted lines indicating the location of the *fanya juu* constructions.

Figure 5.13 shows the pictures obtained on (a) 8 April and (b) 11 April 2006 at Site 4. The *fanya juu* zones clearly show the high moisture zones as a result of ponding at the bunds. Again infiltration potential is shown to be highest at these zones. A continuous link between the high moisture zones at the surface and sub-surface zone would be expected. However, this has not always been the case. This is largely attributed to the weaknesses in the ERT approach taken. A possible existence of zones which promote lateral drainage could explain this break in continuity.

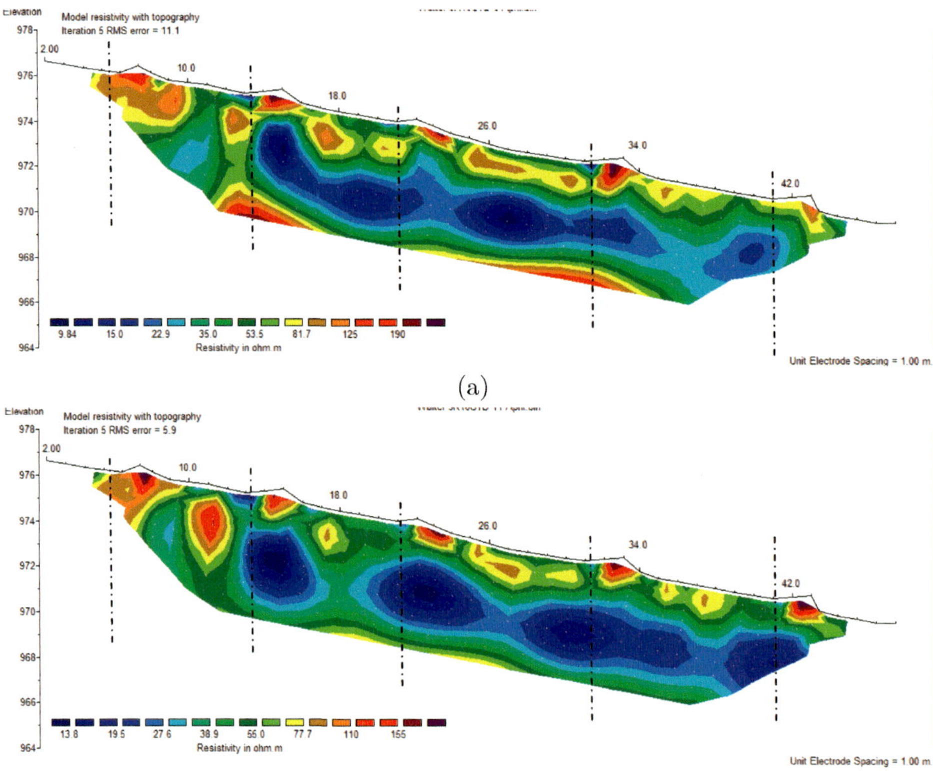

(b)

Figure 5.13 Absolute resistivities [Ω.m] at Site 4 on (a) 8 April and (b) 11 April 2006 at Site 4 with dotted lines indicating the location of the *fanya juu* constructions.

5.4.6 Discussion

This section has demonstrated the application of geophysical techniques to understand soil moisture dynamics. The findings have been compared against repeated measurements at the same sites.

Repeated measurements show that the soil moisture distribution is not uniform in the cultivated strip between the *fanya juu* constructions. At both sites the control tube records the least moisture levels most of the time. The middle tube, Tube C, reflects low soil moisture levels which are similar to the control tube (Tube A) while the downslope tube, Tube D, records the highest moisture levels. The box plots demonstrate that the variance in soil moisture levels is reduced as a result of more water available especially at ponding zones. This reduction in variation leads to a reduction in the extremes caused by dry spells during wet seasons. This is particularly true at Site 3 (Iddi) where more diversion took place.

The ERT pictures confirm the presence of "wet zones" around the location of the *fanya juu* constructions. These wet zones suggest more moisture retention as a result of improved infiltration in the trenches or at the ponded parts. These zones also retain higher moisture levels during drying phases and, most likely, during the dry season as well. This has also been confirmed by the output from the HYDRUS2D model. For the farmer, planting of deep rooting crops can tap into these moisture zone and hence survive even through dry seasons (Makurira et al., 2009a). This is good for all year round crop diversity and, consequently, food security.

While ERT techniques have proven to be applicable in studies of this nature, it is acknowledged that more refined output could have been obtained with better geophysical expertise during the period of data collection. For instance, the dipole-dipole array setting used during the research may not be the most appropriate for the dry conditions prevailing on site. A different array setting, e.g. Pole-Pole, Pole-Dipole or Wenner setting, could have produced better results through more acceptable data sets for analysis (Loke, 2003) due to a better signal to noise ratio. The electrode spacing of 2 m used is also too wide for such as it results in deeper signals while the research is only interested in the upper 2m or so of the soil profile. This results in too coarse a resolution in the upper soil horizons.

5.4.7 Conclusions

The research has demonstrated that the ERT geophysical method can be used to complement detailed soil observations on site. This method can also be applied to shallow depth levels although it is recommended to conduct such research with good background knowledge of geophysical techniques. For a typical research of this nature which focuses on shallow depths, it is recommended to use electrode spacing of 0.5 m (or less). Larger electrode spacing implies investigating to deeper depths but with lower resolution (Wenninger et al., 2008).

The geophysical technique is most suited where rapid assessment is required and where soil disturbance is to be minimised as much as possible. Also, since the method does not give direct soil moisture values, it can only be applied where indicative soil moisture changes suffice. This method has the potential to clearly map the heterogeneity in the soil material which is created by the in-field interventions with the purpose of deliberately altering the field scale water balance.

Chapter 6

PRODUCTIVITY ANALYSIS[5]

6.1 INTRODUCTION

Grain yields have been noted to oscillate below 1 t ha^{-1} in many sub-Saharan smallholder rainfed farming systems (Bhatt et al., 2006; Rockström et al., 2007; Rockström et al., 2004), yet there is potential to double or even quadruple current yields (Rockström et al., 2007) if soil moisture retention can be enhanced as a mitigation strategy against the impacts of dry spells. A common measure of the efficiency of farming systems is the obtained yield per season. For the common farmer, this can simply mean the number of grain bags obtained at the end of the season. If more harvests are required, the solution would be to simply increase the land under cultivation. This approach is more applicable where land and water resources are not limited. However, under the existing challenges of water scarcity in

[5] Based on the following papers:

c) Makurira, H., Savenije, H.H.G. and Uhlenbrook, S., 2007b. Towards a better understanding of water partitioning processes for improved smallholder rainfed agricultural systems: A case study of Makanya catchment, Tanzania. Physics and Chemistry of the Earth, 32(15-18): 1082-1089.

d) Makurira, H., Savenije, H.H.G., Uhlenbrook, S., Rockström, J. and Senzanje, A., 2009b. Investigating the water balance of on-farm techniques for improved crop productivity in rainfed systems: A case study of Makanya catchment, Tanzania. Physics and Chemistry of the Earth, 34: 93-98.

the sub-Saharan region water productivity would be a more appropriate measure of the efficient utilisation of cropping systems. Water productivity (WP) measures the biomass produced from a unit of water or, inversely, WP is a measure of the amount of water required to produce a given amount of biomass. Where WP values are extremely low, it may be more efficient to import rather than to cultivate crops in areas where the opportunity costs of using water for other purposes are very high (Dabrowski et al., 2009).

This chapter analyses the impact of the tested techniques on water productivity levels at farm scale.

6.2 Data collection

Daily rainfall, runoff and yield data were obtained as explained in Chapter 4. The obtained yields for each treatment were calculated by measuring the dry weight of grains according to the different treatments tested in the research.

6.3 Data analysis

The rainfall received at each site has been noted. Further, the additional water obtained through runoff diversions has been estimated from the tipping bucket measurements. This gives an estimate of the total water available at each site to support crop growth. This water availability, when analysed together with the grain yields obtained, leads to calculation of water use efficiencies within the field.

The grain yield produced (kg ha^{-1}) at each site and for each treatment has been obtained by measuring the mass of grain obtained from a measured cultivated measured. The maize yield has been analysed according to season, treatment and site. This allows for the comparison of different treatments according to the location and for each season. When compared against the total seasonal rainfall received, the productivity of each treatment can be determined. The water productivity (WP) (kg m^{-3}) has been calculated as the grain yield (kg) obtained from a unit of water.

The student t-test for comparison of means has been applied to test if there is significant differences ($p<0.05$) in the means of the calculated yields. This method of analysis applies where the distribution of data follows a normal curve.

Water availability

The seasonal variations in rainfall are high. Table 6.1 summarises what is typical for the given hydro-climatological conditions. Season 2 received the highest average rainfall (484 mm/season) while season 3 recorded the least average rainfall (108 mm/season). The contribution of runoff from total water availability varied for the different sites and seasons.

Table 6.1 Seasonal water availability per site during the period 2006-2007.

	Site 1			Site 2			Site 3			Site 4		
	P (mm)	Q (mm)	P+Q (mm)	P (mm)	Q (mm)	P+Q (mm)	P (mm)	Q (mm)	P+Q (mm)	P (mm)	Q (mm)	P+Q (mm)
Season 1	303	21	324	399	40	439	203	281	484	359	137	496
Season 2	600	42	642	592	59	652	308	388	696	437	127	564
Season 3	121	8	130	123	12	135	113	147	260	105	42	147
Season 4	258	18	276	299	30	329	181	201	382	148	56	201

Where (all terms in mm/season):

P: seasonal rainfall; Q: contribution from runoff diverted onto the field; P+Q: total seasonal water availability

6.3.1 Grain yield

The maize grain yields obtained have been compared for each treatment and according to site.

Figure 6.1 shows the average grain yield obtained for each site over the four measured seasons. Season 1 and Season 2 received near to average seasonal rainfall with a combined average rainfall of 316 mm/season and 484 mm/season, respectively. With diversions, the average water availability increased to 436 mm/season and 638 mm/season during Season 1 and Season 2, respectively. On the other hand, seasonal rainfall was very low during Season 3 (average of 115 mm/season). Although slightly higher rainfall was received in Season 4 (222 mm/season) the distribution of the rainfall was very poor (Figure 4.8). As a result, harvests of above 1 t ha^{-1} were obtained during the first two seasons while Season 3 was a complete crop failure at all sites. The diversions helped to obtain small yields at Site 3 and Site 4 during Season 4.

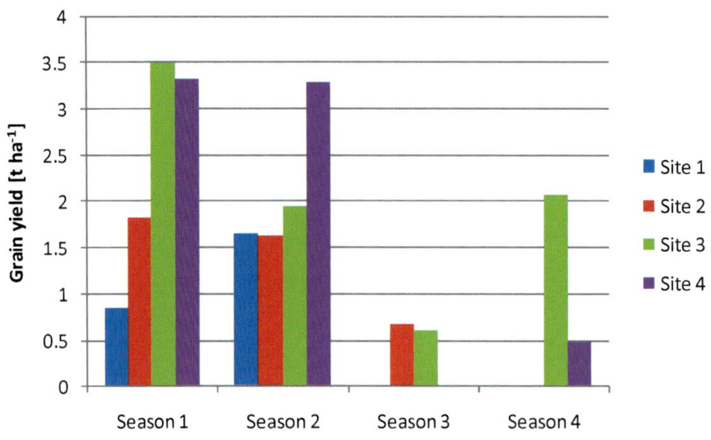

Figure 6.1 Average seasonal yields at each site over four seasons (2006-2007).

Figure 6.2 shows the averaged performance of each treatment over the four seasons. At three of the four sites studied, the control plots (traditional cultivation practices) recorded average yields of less than 1 t ha^{-1} (except at Site 4). Site 3 and Site 4, which adopted the introduced techniques more comprehensively, recorded higher yields with Treatment 2 (ripping, manure and runoff diversion) recording the highest yields, on average. At Site 2, Treatment 2 also included a cover crop of beans (Enfors et al., 2010; Enfors et al., in revision).

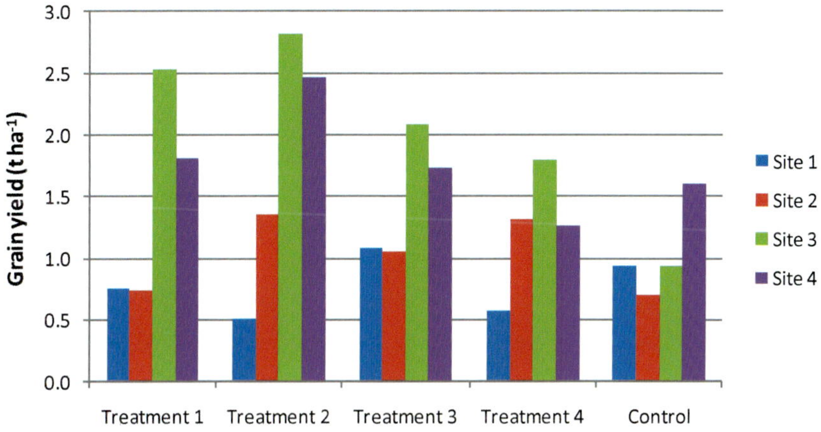

Figure 6.2 Average grain yields from four seasons per treatment for the four study sites over four seasons (2006-2007).

At Site 3 and Site 4, for the last season, grain yield per crop line was also measured along rip lines across the cultivated *fanya juu* terraces for Treatment 1. This provides an indication of the impact of the in-field interventions on the distribution of productivity with distance from the trenches or bunds. In agronomical analysis, it is normal to discard the rows at the edges as they are usually affected by external factors. In the case of Figure 6.3 row 1 and row 11 for case (a) and row 1 and row 9 (for case (b) can be ignored. It can be seen from Figure 6.3 that yields are higher in the rows closest to the downslope *fanya juu*. This can be attributed to the effect of ponded water around the bunds that infiltrates and helps to increase soil moisture availability (c.f. Chapter 4).

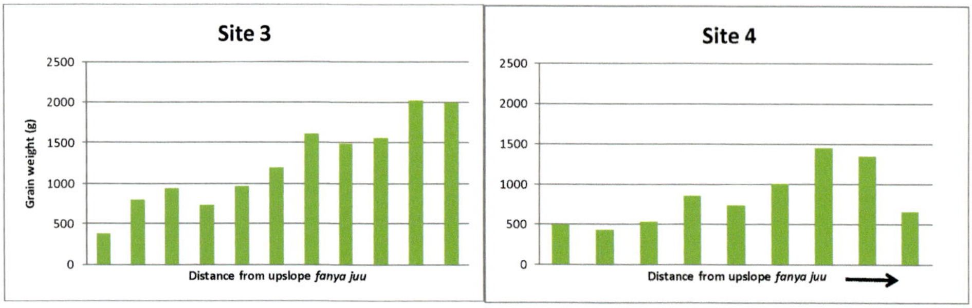

(a) (b)

Figure 6.3 Distribution of grain yield per rip line as measured from upslope of terrace at (a) Site 3 and (b) Site 4 in the *Vuli* 2008 season.

Statistical analysis

The yield data obtained did not fit a normal distribution as required for the t-test for statistical comparison of means. The yield data only attained a normal distribution after logarithmic transformation after which the t-test analysis applied (c.f. Enfors et al., 2010).

From the yield data obtained at all four sites, the statistical analysis showed significant differences ($p<0.05$) in yields due to seasonal variations and location. This is logical given the large variation in seasonal rainfall received. It was not possible to carry out a statistical test for significance of differences in treatment at each site due to the small sample size of 4 seasons measured at each site.

6.3.2 Water productivity

Water productivity is calculated as the total amount of water available to the crop (from Table 6.1) which produced the grain yields (Table 4.1). It is observed that water productivity increased from an average of 0.31 kg m^{-3} in Treatment 4 to 0.45 kg m^{-3} for Treatment 2. The average productivity values in the control have been distorted by uncharacteristic high yield values in the control plot at Site 4.

Table 6.2 Average water productivity for all four sites with different treatment during the study period (2006 – 2008).

Site	Water productivity (kg m^{-3})				
	Treatment 1	Treatment 2	Treatment 3	Treatment 4	Control
1	0.22	0.15	0.32	0.17	0.29
2	0.19	0.35	0.27	0.34	0.20
3	0.55	0.62	0.46	0.39	0.37
4	0.51	0.70	0.49	0.36	0.61
Mean	0.37	0.45	0.38	0.31	0.37

Figure 6.4 was then plotted to find the relationship between the calculated productivity and grain yields. A linear relationship between water productivity (kg m^{-3}) and grain yield of the crop (kg ha^{-1} $crop^{-1}$) was obtained which shows the level of productivity for the grain yields realised. The inverse of the slope of the graph gives the crop water requirement under the studied conditions. The data points shown for each treatment are bound by the "perfect yield" line of 450 mm/crop. This line is drawn based on maize seasonal water requirements calculated for the study area as reported in Makurira et al. (2007a). The lowest yields have the largest distance to the line. A few inconsistencies in the treatments occurred which can be attributed to the seasonal distribution of dry spells and different on-site management approaches. In general, the effect of diverting water has the biggest effect on the observed yields.

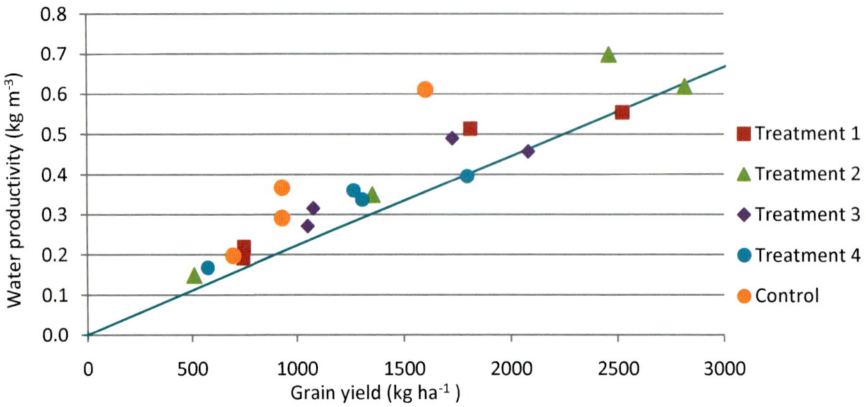

Figure 6.4 Variation of total water productivity with grain yield at all sites and seasons (2006 - 2008).

6.4 Discussion of results

6.4.1 Water availability

The total rainfall received during the research was always less than 500 mm/season. This suggests that the rainfall received in the area is insufficient to satisfy staple food seasonal crop water requirements for maize. The seasonal rainfall amounts were highly variable across the research sites. Diversions were most effective at Site 3 and Site 4 where, even with the least rainfall, the water availability was much higher. In this case, Site 3 supplemented, through diversion, 109% of available rainfall compared to Site 4 (about 50%). The diversions were much less effective at Site 1 and Site 2 where, due to very little diversion efforts, only less than 10% of the available rainfall could be supplemented. If water is accepted as one of the key limiting factors in crop productivity then the results obtained suggest that it is most important to focus on diverting as much water as possible onto field plots as one important element of reducing the impact of insufficient rainfall and dry spells. The high variability of already deficient rainfall highlights the need for interventions to cushion crops against mid-season dry spells.

6.4.2 Yields and water productivity

The yields obtained confirm that, in the absence of improved farming techniques, crop yields of less than 1000 kg ha^{-1} will be realised as shown by the harvests in all of the control plots. The higher yields obtained at the control of Site 4 (Figure 6.2) can be explained by the fact that the soils at the control section at Site 4 have a higher water holding capacity especially in the drier season (Figure 5.9 and Figure 5.11) and, hence, offer more resilience against dry spells. Also, during Season 1 and Season 2, the rainfall received was more than 300 mm/season which, when harvested provided reasonable amounts to support crop growth especially when the seasonal distribution of the rainfall is good. This implies that the effect of the improved techniques is hardly noticeable when there is sufficient water. Similarly, extremely dry seasons (e.g. Season 3), the techniques also fails as there is nothing to divert. However, when there is just insufficient water to meet crop water requirements, these techniques play a significant role in obtaining at least some harvests.

The variability of yields for the different treatments shows that Treatment 2 (ripping, manure and runoff diversion) produced higher yields than the other treatments. The higher yields of more than 1500 kg ha^{-1} obtained at Sites 3 and 4 in average seasons confirm that there is scope to triple or even quadruple yields, if more efficient techniques were employed (Rockström et al., 2007). The distribution of yields by rip line suggests that higher productivity occurs just uphill of the bunded section of the interventions where ponding occurs. This ponding effect promotes infiltration, better soil moisture availability and deposition of nutrient-rich fine sediment. This fine sediment also improves the water holding capacity in sandy soils. However, in the rare good rainfall seasons, it is also possible that these ponding zones create water logging conditions leading to yield reductions.

The average water productivity improved with the applied innovations by 0.35 - 0.51 kg m^{-3} and, for the improved trials, the result obtained are comparable with observations in irrigated schemes (i.e. 0.4 - 0.7 kg m^{-3}) under comparable environmental conditions (Igbadun et al., 2006).

The statistical analysis did not show significant differences in yields due to different treatments. Enfors et al. (in revision) and Munodawafa and Zhou (2008) made similar observations on different trials in Tanzania and Zimbabwe, respectively. However, the absence of significant difference should not be misinterpreted to suggest that the "improved" techniques do not have meaningful impacts on farming systems since it may take a long time before noticeable impacts of improved can be observed on the ground.

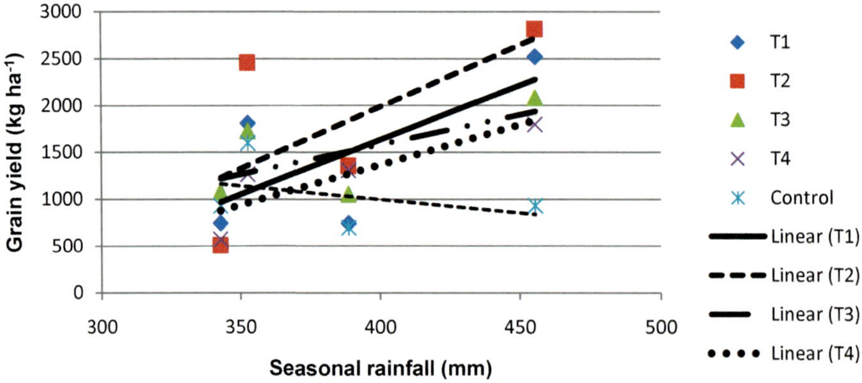

Figure 6.5 Response of yield per treatment to water availability as observed in all seasons (2006 – 2008).

Figure 6.5 shows the clear difference in responses to yield obtained from different treatments in response to water availability. There is better potential to obtain higher yields under improved tillage techniques than with the traditional farming practices. This potential is boosted (for example in Treatment 2) where a combination of water and nutrient management are applied. In reality, a steeper positive gradient of the control graph would be expected but, in this case, this has been distorted by unusual high yields in the Control at Site 4 during the first two seasons.

6.4.3 Additional benefits of "improved" techniques

The *fanya juu* techniques which were tested in this research provided not only improved grain yields at field scale, but also altered the soil moisture retention characteristics (Makurira et al., 2009b). The high infiltration zones created by the *fanya juu* technique retain moisture for longer periods and farmers can take advantage of this situation by growing fruit trees, grass for cattle feed and other diversified crops throughout the year. As a result, even in dry seasons, farmers now harvest some crops as shown in Figure 6.6 where a farmer (Iddi, farmer of Site 3) harvested the largest cassava root ever seen in the area.

Figure 6.6 Cassava harvest in dry season at Site 3.

6.5 CONCLUSIONS

This chapter confirms that water availability during the cropping season is a major challenge, particularly where traditions constrain farmers to continue growing relatively high water demanding crops in areas of little rainfall. The research also confirmed the scope for obtaining improved yields provided more efficient soil and water conservation methods are applied. Highest yields can be obtained where a combination of techniques are applied, which shows that there is no single solution to the challenges faced by resource-poor, rainfed farmers. In this research, a combination of diversion, ripping, manure and cover cropping has shown the greatest potential to result in better yields.

The *fanya juu* technique in combination with diversions has proved to be effective in boosting yields. This is an additional advantage to the commonly understood soil conservation functioning. This is important if it is agreed that water is the limiting factor in rainfed farming systems. The fact that the soil moisture distribution pattern is altered is a major advantage in that even with little water harvested, the wetter zones can better support crop growth. There is, however, the higher risk of water logging in the wetter zones especially under extremely wet conditions. However, in such wet conditions, the elevated bunds would offer excellent growing conditions thus creating flexibility in growing conditions and more resilience against extreme climatic conditions. For areas with higher potential for runoff, and where the terrain permits, it may be advisable therefore to space the *fanya juus* as close as possible, particularly since the *fanya juu* technique does not result in significant loss of cropping area when alternative cropping opportunities are explored. The farmer can grow bananas, cassava, paw paws and more perennial crops around the *fanya juu* structures.

While the techniques tested show the potential to significantly improve yields in average seasons, it has been shown that the effectiveness of these techniques is not so significant in good seasons where crop water requirements can be widely satisfied with rainfall alone. The techniques are also not effective in extremely dry seasons where total crop failure was noted. Consequently, an important conclusion that can be drawn from the current research results is that these techniques are most efficient in average seasons and where the possibility of runoff diversion exists. They are also very effective in mitigating the effects of long dry spells during crop growth stages as, in farming systems, the temporal distribution of water availability is important in addition to the total seasonal water availability. Comparison of the performance of different treatments was only possible where sufficient runoff diversion was possible which enhances the argument that water is the limiting factor to crop productivity in these farming systems.

Finally, it can be concluded that the success of the tested "improved" techniques should be looked at from a wider perspective where, in addition to more than doubling the current yield levels under traditional farming systems, farmers have the option to practice dry season cultivation of alternative crops which increase food security levels throughout the year. This food security is also complimented by better diets which are derived from a larger diversity of crops.

Research of this nature leads to more questions. For instance, what is the downstream implication of local successes in converting generated water to green water processes? It is necessary to conduct more comprehensive assessments of the impacts of successful uptake of such "improved" techniques on the overall catchment hydrology including processes such as flood generation, erosion, groundwater recharge, water availability downstream and different evaporation fluxes. It is noted that transpiration fluxes increase with the tested techniques but, in the absence of storage structures, deep percolation also increases and accounts for most of the available water. In that case, while productivity increases, the water "lost" to deep percolation is, in fact, recharging groundwater and may become immediately available to the local farmer or to downstream ecosystems through sub-surface flow. This would be another positive benefit from these techniques when looked at from a larger scale perspective.

Chapter 7

SYNTHESIS OF THE RESEARCH

This research has highlighted the challenges faced by smallholder rainfed subsistence farmers in the studied area in the sub-Saharan region where several factors (e.g. degrading soils, lack of resources to invest in farming systems, social and cultural practices and harsh climatological conditions) combine to result in low yield levels for the common crops which are grown in the region. Maize productivity has been studied in this research because maize is the most popular staple food crop in the region. It is assumed that other crops would behave in a similar manner under the same conditions.

The fact that maize yields (and most likely that of all other grain crops) have not improved in the last three decades (FAO, 2005) in a region where the rate of population growth is among the highest in the world (UNFPA, 2008) shows that there is an increasing likelihood that food deficits and, hence, imports will occur more frequently to avert famine disasters (de Fraiture et al., 2007; IWMI, 2003). In this research, the rainbow of water concept (Savenije, 2000) has been adopted where water production for agriculture is a transpiration (green water) issue. The optimisation of green water flows at the expense of "blue" and "white" water fluxes leads to increased transpiration, hence, biomass production. The conversion of "blue" water to "green" through rainwater harvesting is one way of increasing productivity but increasing pressures on "blue" water through scarcity and competition makes it more appropriate to focus more on the conversion of unproductive "white water" fluxes (evaporation) to productive "green water" fluxes.

This research has explored alternative options for coping with low yield levels under the existing challenges. Whereas many studies have attempted to contribute towards improved productivity through a shift from the traditional practices (Barron and Rockström, 2003; Kosgei et al., 2007; Ngigi et al., 2007; Temesgen et al., 2007), the link between promising techniques and the likely hydrological consequences to immediate and downstream ecosystems is often less understood (Jewitt, 2006; Rockström et al., 2004). This research has also attempted to bridge that gap by analysing the tested techniques from a multi-disciplinary perspective. The response of maize yield with treatment and available water, for instance, is a typical focus area for agronomists. Hydrologists, on the other hand, would have been more interested in the partitioning processes of the available water. In between, the soil and water dynamics at field scale have been explored in detail using different approaches (e.g. TDR, ERT and HYDRUS2D modelling). Uhlenbrook (2007) acknowledges that different experimental techniques and hydrological modelling approaches need to be developed to further answer questions related to the interactions between rainfall, runoff and agriculture. In the end, results from all the analytical approaches confirmed the effectiveness of the tested techniques in providing scope for improving yields obtained by rainfed farmers under harsh climatic conditions.

In selecting techniques for testing, consideration has been given to simple and affordable solutions which can easily be tried and adopted by local communities. The research articulates the following key components of upgrading rainfed agriculture in semi-arid environments:

a. Water availability for agriculture

Comprehensive on-site measurements have been carried out during the research to better understand the hydro-meteorological characteristics applicable at field scale and then to translate these into soil and water relationships which, ultimately, influence crop performance.

The rainfall received of less than 400 mm/season against high potential evaporation and transpiration rates of up to 10 mm d^{-1} (Mul, 2009) imply that the water available to the root zone, after evaporation and runoff processes, is hardly enough to support any crop to maturity. Yet, if all this rainfall were to infiltrate into the root zone, then yields of more than 1 t ha^{-1} can be realised. Alternatively, the deficit in rainfall can be met by rainwater harvesting and supplemental irrigation. In this research, runoff from gullies and undefined flow paths has been diverted onto the field plots to counter moisture deficits through insufficient rainfall, evaporation and runoff losses.

b. Improvements in cultivation techniques

Seed management

Selection of appropriate seed is one of the basic steps that are necessary to promote better yields (Nyagumbo et al., 2009). This ranges from the purchase of treated and appropriate varieties for particular geographical environments to the selection of good seed after harvesting in cases where seed purchasing is unaffordable. While this research has focused on soil and water relationships but parallel efforts on seed breeding are underway. Research in appropriate seed varieties is required to match the research in water for agriculture (Kijne et al., 2009).

Soil and water conservation

Another major challenge to cropping systems is the continuous loss of soil and nutrients through erosion. The *fanya juu* technique was originally advanced as a soil conservation measure in steep slopes (Bewket and Sterk, 2002; Gichuki, 2000; Tenge et al., 2005) but the benefits to crop productivity through improved soil and moisture retention are now being increasingly recognised (Makurira et al., 2007b; Motsi et al., 2004; Mwangi et al., 2001; Nyagumbo et al., 2009). The modified *fanya juu* technique allows water to cascade within the field after filling the trenches within the cultivated field. This way, the technique performs the dual purpose of reducing runoff velocity hence reduce soil and, hence, nutrient loss while, at the same time, retaining more water through in-field temporary storage. In addition, the technique favours percolation to the groundwater reserves at the expense of surface runoff. Through this research, it has been shown that the *fanya juu* technique changes the moisture distribution across the cultivated field leading to higher moisture zones around the trenches and bunds. These high moisture zones facilitate the growing of longer season crops and also provide higher moisture levels to counter the impacts of dry spells. In extremely wet seasons, however, the ponding zones may create water logging conditions. The *fanya juu* technique performs well at sites with moderate slopes (<20%) (Bewket and Sterk, 2002). The spacing between the trenches/bunds is dependent on the slope with shorter spacing being ideal for steeper slopes.

The *fanya juu* technique has therefore been shown to be efficient in both soil and water conservation which offers food security in both wet and dry seasons. In extremely wet seasons, the ponding zones create water logging conditions which means that crops may not perform very well. However, the otherwise drier parts of the field, corresponding to locations of Tubes B and C, would attain sufficient moisture levels to support crops. The bunds crops on the bunds also do well since the bunds are not affected by water logging. On the other hand, in dry seasons, crops around Tube C may not do so well but the ponding zones contain extra moisture which supports crops. The crops on the bunds always contain sufficient moisture levels to survive. This therefore means that the *fanya juu* technique provides enhanced security of harvests in both very wet and just below average rainfall seasons. This is an excellent case of building resilience at field scale as highlighted by Enfors and Gordon (2007; 2008).

Conservation tillage

The Magoye ripper has been used during land preparation. The ripper cuts along planting lines with deep and sharp penetrations. The ripping technique concentrates soil infiltration along the planting line. In the process, the minimum disturbance of the general soil structure results in less soil evaporation. The ripper also breaks the hard pan which is associated with the use of traditional implements such as the plough and hand-hoe. While the ripper is efficient from a water conservation perspective and that it is time efficient, it has disadvantages in weed control where more weeding events would be required during cropping seasons.

c. Yield analysis

The introduced techniques have resulted in yield levels of up to 4.8 t ha^{-1} being observed at the study sites against yield levels of less than 1 t ha^{-1} under traditional farming systems (Makurira et al., 2007b). Diversion of runoff alone resulted in an increase in average maize yields from 0.9 t ha^{-1} to 1.4 t ha^{-1}. The average maize yields further increased to 2 t ha^{-1} when manure and cover cropping were introduced. This shows that water availability plays a key role in improving yield levels. However, further improvements can be realised if nutrient management is balanced with water availability. Only organic manure, which is cheap and mostly available, has been used in this research. The nutrient content of this manure is largely dependent on the quality of food available to the animals. This is highly compromised in degraded and water-scarce regions like the study area. This suggests that, if organic commercial fertilisers were used instead, the yield levels realised would be much higher. In addition, the use of pesticides and fungicides helps to control yield reduction. However, communities would need to be conscientised on health and safety considerations when applying chemicals in the field.

The variation of maize yield with distance from the trenches shows the functioning of the *fanya juus* in the overall distribution of soil moisture within the cultivated field. The ponding zones create areas of higher moisture concentrations, hence higher yields, in average seasons. In wet seasons, these zones may create logging conditions leading to crop failure. In such instances, the otherwise drier areas, would have sufficient moisture to support crop growth. On the other hand, alternative cropping such as bananas, pawpaws, cassava and fodder compensates for the loss of land under maize cultivation. These alternative crops offer more protection against climate change and climate variability hence improve on food security as they can also be harvested in the dry seasons.

d. Water partitioning

The seasonal rainfall received in the study area is generally insufficient to support crops to maturity. Rainfall and runoff events are of short duration and, if not converted to immediate beneficial use, the runoff contributes to river flows or recharges groundwater reserves along river channels. Rainwater harvesting captures this rainfall as well as the runoff generated in the micro-catchment. In average seasons, diversions from gullies, road drainage and not so distinct flow paths help to concentrate water into cultivated areas and supplement the, otherwise, insufficient rainfall in providing water for crop growth (Hatibu et al., 2006). This is a simple and cheap way of enhancing water availability within cultivated areas.

In this research, increased crop productivity is viewed from the perspective of altering the current water partitioning where unproductive processes (i.e. soil evaporation, interception and runoff) account for more of the available water at the expense of "green" water for transpiration purposes. The techniques that have been tested in this research aim to alter these partitioning processes to divert more water to transpiration. More water diverted into the root zone increases the potential for transpiration to occur.

Direct and indirect methods have been used to better understand these partitioning processes. The results show that the amount of water available for diversion is dependent on the size of the runoff generating catchment. Larger catchment areas have more diversion potential. Ultimately, water security and, hence, more productivity is better achieved with increasing diversion potential.

An analysis on water partitioning processes at two sites over four seasons shows that the diversion increased water availability by 36% and 129%, respectively, depending on the diversion potential. In the process, green water flows (transpiration) increased by 40% and 57%. However, more diversions also resulted in more groundwater recharge and this accounted for 42 - 47% of the available water. High evaporation levels imply that a significant proportion of the available water is attributed to unproductive purposes. If part of this water were to be diverted to transpiration purposes then the obtained yields could further improve.

This partitioning analysis leads to a number of interesting conclusions; first, the transpiration flux increases with water availability until the stage when all crop water requirements are met. Secondly, from the rainfall and runoff patterns, it is clear that the techniques used cannot absorb all the harvested water hence the increase in groundwater recharge with more water diverted. This means that storage structures would be required to store this excess water and use it as supplemental irrigation when dry spells occur. From a whole catchment perspective, however, another conclusion from the partitioning analysis would be that the groundwater recharge is an indication of more infiltration and, hence, longer season crops can be grown around the trenches and bunds. This excess water recharges the groundwater system and eventually benefits downstream ecosystems.

e. The challenges of adoption and adaption

This research has demonstrate that there is indeed scope for improved productivity in smallholder rainfed systems through the adoption of more efficient soil and water conservation techniques. Many studies have been concluded with success stories demonstrating better scope for improved productivity, yet the situation on the ground has not changed much (FAO, 2005). It is therefore necessary to reflect on why successful studies are not immediately converted to success stories in the field through the adoption of tested techniques.

Project 1 of the SSI programme has explored this issue in detail. Indications are that adoption is a function of social capital, prior information availed to beneficiaries and potential for improved incomes, amongst others (Tumbo et al., 2010). In general, it has been noted that research which is divorced from the targeted beneficiaries will not likely be adopted despite promises to change livelihoods. As a result, the SSI programme took a participatory research approach where the beneficiary farmers were involved in the research from as early as possible until the end. That way a sense of ownership and partnership is cultivated between the research and the beneficiary of the research. The need for a participatory research has been emphasised by Mirghani and Savenije (1995) and Bewket and Sterk (2002).

The other issue which is important to the successful adoption of promising strategies is the need to appreciate what is achievable within the target group. Whereas many possible solutions exist to solve the challenges in the study area, this research focused on solutions which are easy to understand, are affordable and hence are implementable at ordinary household levels. Incremental yield benefits are linked to investment costs (Nyagumbo et al., 2009). However, since the majority of smallholder farmers are resource poor, they are only restricted to techniques which provide additional benefits only up to what they can afford to invest. Affordable investments often relate to in-field solutions. Stream diversions, *fanya juu* constructions, deep tillage and nutrient management are such solutions which do not require much capital investment but, if managed properly, have the potential to more than double yields (Kijne et al., 2009). This has been investigated and proved in this research.

Adoption can only be possible if the farmer can build on previous seasons' success to further invest in the agricultural activity. If the farmer is not guaranteed of a good return to their investment then they are not willing to gamble the little resources at their disposal. The sustainability of projects is related to the perceived tangible benefits to the people (Vishnudas, 2006). If the people cannot be convinced of any tangible benefits then their willingness to adopt is highly reduced. Figure 4.1 showed the link between investment costs, techniques applied and the resultant yield responses. Figure 7.1 shows the dynamics of investment and returns in a farming enterprise when expected benefits at each stage are brought into the picture.

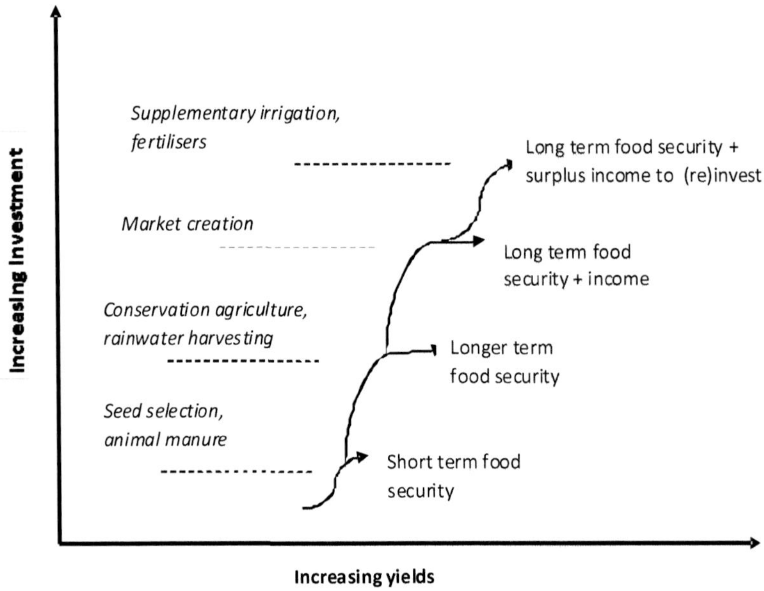

Figure 7.1 The link between investments, yields and derived benefits.

In this figure, external investment can be any of the techniques discussed in this thesis or can be other investments in irrigation infrastructure. However, without markets for the produce, the farmer is discouraged to produce more as he has limited options for excess produce. On the other hand, if markets exist, additional income can be obtained which can either be re-invested into farming or, alternatively, pay for other household priorities.

f. Policy issues

There is need to link research, governance and on the ground conditions. Not much success is achieved where these three factors do not converge (e.g. van der Zaag, 2005). This research has demonstrated good linkage between the research itself and the participation of the beneficiaries of the research. The third dimension to the relationship is for research findings to influence policy. Many rural communities find it easier to adopt practices which are promoted by their governments. If policy makers are involved in research and appreciate the research findings, then it would be easier for adoption through government promotion.

g. More multi-disciplinary and integrated research

The SSI programme has succeeded in articulating the challenges faced by rainfed subsistence farmers in semi-arid sub-Saharan Africa through a multi-disciplinary approach. Kongo (2008) and Mul (2009) have explored the dynamics of larger scale water processes in contributing to water uses at local scales in Tanzania and South Africa, respectively. Both studies acknowledge the absence of sufficient data to facilitate decision-making hence the need to install as many measuring devices as possible. Kosgei (2009), based on research in the Thukela Catchment in South Africa, also demonstrated that there is scope to improve productivity levels through better management of available water, including investment in sub-surface storage structures. Enfors (2009), on the other hand, acknowledges that any improved techniques that are tested will take a considerable time to change the existing soil fabric to be able to cope with climatic shocks. Komakech et al. (submitted) (submitted) and Kemerink (2009) demonstrate that local water sharing arrangements between competing users are key to the success of sharing scarce water resources and optimising benefits from the little available water. However, at the end of the day, it is necessary to explore why, even despite the potential benefits of adopting the promising tested techniques, the uptake of these innovations by local communities is usually below expectations. Adoption and adaption are also a function of how new ideas are sold to local communities and the perceived benefits to be derived from adopting such innovations (Tumbo et al., 2010).

From the above, it can be deduced that the scope for improving livelihoods exists and is a function of socio-economic, ecological, hydrological and agronomic considerations. More research of this integrated nature is needed to stimulate attention towards where the greatest mileage for food security can be covered i.e. in rainfed farming systems.

Summary of synthesis

Figure 7.2 shows an experimental overview consisting of a typical research site. At the site the natural system is shown together with alterations to the natural system such as diversions and *fanya juu* constructions. Typical observation locations such as a rain gauge, runoff measuring stations and locations for soil moisture monitoring are also shown within the cultivated plot.

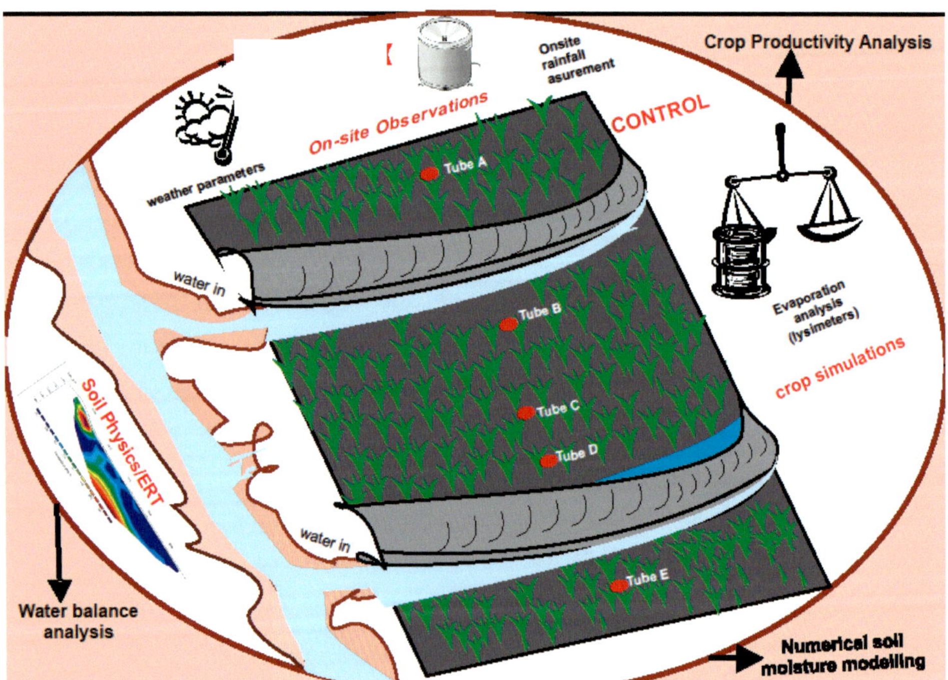

Figure 7.2 Overview of the synthesis of the research.

From such an experimental setting data has been collected to answer the research objectives using a converging evidence approach.

Table 7.1 shows a summary of techniques that have been applied in this research to better understand soil and water interaction at field scale. The figure shows that these techniques complement each other to confirm impacts of important controlling variables which have been analysed in this research. The fact that each controlling variable is investigated by more than one technique shows that the research applied different methods to address the research objectives. After applying the different techniques, the results obtained confirm similar findings which have led to better understanding of partitioning processes at field scale. In this way, it has been possible to understand water productivity for the maize crop and to arrive at firm conclusions that the scope to improve water productivity exists if only famers can re-partition the water balance at field scale through the adoption of more efficient farming methods.

Table 7.1 Summary of techniques applied and their relationships with the main controlling variables.

Technique / Controlling variables	On-site observations *(e.g. rainfall, runoff, biomass, grain yield, etc)*	TDR	ERT	Spreadsheet modelling	HYDRUS2D
Diversions	×	×	×	×	×
Ponding	×	×	×	×	×
Slope	×	×	×		×
Moisture retention	×		×	×	×
Soil texture	×		×	×	×

Chapter 8

Conclusions

This thesis has detailed the challenges faced by sub–Saharan African (SSA) rainfed farmers who form the vast majority of the population in the region and are faced with the prospects of failing to meet the MDGs and, particularly, those related to hunger and poverty alleviation. Grain yields have not improved in the last three decades yet the region also witnesses the highest population growth in the world. This thesis has successfully demonstrated that, despite the challenges of sub-optimal rainfall for cropping seasons, dry spell occurrences and increasing soil and nutrient loss, the low yields currently obtained in semi-arid SSA environments can be improved at least two-fold through the repartitioning of current moisture processes at field scale. This can be achieved through a shift from the current cultivation practices to the use of innovative techniques at farm level. In practice, this means redrawing the pattern of the "rainbow of water" to result in the conversion of more "blue" and "white" water to "green" water fluxes.

The research has tested a combination of techniques which are believed to be affordable and, yet, are efficient enough to boost yields. Since water is a major limiting factor to crop productivity, any possible means to capture the little available rainfall, when it occurs, should be pursued. Additional water in the field, through diversions, helps reduce the impact of frequent dry spells which affect yields. When water is diverted, soil moisture within the season has been observed to reach suitable levels for crop growth much sooner than without diversions. This means that the diversions help to increase the length of the growing season while smoothing the impact of mid-season dry spells. In the research, the addition of water alone resulted in average maize yield increases of 12.5%. The most efficient treatment (Treatment 2) which includes conservation agriculture, application of manure and cover cropping improved yields by an average of 76%. While the benefits derived could not be quantified according to each treatment (e.g. diversion, ripping, *fanya juus*), it is clear that no single solution exists to solve these challenges. A combination of appropriate techniques would be required for each site to improve on the current trends.

The repartitioning of water using the techniques that have been tested in this research shows that the bulk of water that has been diverted is accounted for as deep percolation. This shows that the soil conditions within the field have not changed during the research period to attain sufficient capacity to absorb all the diverted water. To be able to utilise all the diverted water within the field, the soil moisture holding capacity would need to be increased through, e.g. mulching, application of manure and/or the deposition of finer sediment within the field. This process takes longer than the four years of research although indications of soil property changes were beginning to be noticed (Enfors et al., in revision). At the moment, the increased percolation would serve downstream ecosystems through groundwater flow and storage. For the runoff generating areas, however, this water, if stored, can be used during dry periods as supplemental irrigation. Micro-storage structures should therefore be considered as a rainwater harvesting intervention to temporarily store water immediately after runoff events and use it later when dry spells occur.

The *fanya juu* technique has multiple functions at field scale and should be tried in many SSA environments. The technique helps to arrest soil and nutrient loss while performing critical functions in repartitioning the water balance at field scale. The research has shown that, through the functioning of the *fanya juus*, more water retention and enhanced infiltration can be achieved especially in the trenches and at the bunds. In the process more water is availed to the root zone which facilitates more transpiration. However, without diversions, the *fanya juu* technique would not achieve any significant additional benefits.

The variation in soil moisture levels across the cultivated strip creates different moisture conditions in extreme (wet and dry) conditions. In this study, it has been shown that the ponding zones (around Tube D) create above average moisture conditions which result in at least a harvest even in below average seasons. In wet seasons, however, water logging conditions are likely to occur in these zones leading to crop damage. However, in such circumstances, the otherwise drier areas (Tubes B and C) may, in fact, record higher moisture levels to support crop growth. In addition, in either case, crops can also be grown on the bunds. While this arrangement is not good for optimal yields from the primary crop through reduction in land available for planting the main crop, the techniques offer higher security against the impacts of dry spells, hence, more resilience against harsh climatic challenges.

In practice, these techniques create deliberate heterogeneous conditions within the field which function to perform differently under similar hydrological conditions within the same season. If, indeed, poverty and hunger are closely related, then the adoption of these techniques help to improve food security thus helping the farmer to escape from the poverty cycle. It may be argued, however, that this success comes at a cost to productive land within the field. This may be true, to an extent, but the number of seasons when the farmer is guaranteed of, at least, a harvest is increased through the adoption of these techniques.

The success of the *fanya juu* technique that has been demonstrated in this research shows that the farmer has the capacity to alter the soil and water relationships within the field. The technique promotes water retention within the field while, at the same time, facilitating the deposition of nutrient-rich fine sediment which would otherwise be washed away with runoff. This fine sediment enhances soil fertility and water holding capacity. Besides altering the in-field water partitioning processes, the farmer can build on household level resilience by diversifying to other crops especially those which would, otherwise, not grow in the dry season. In this research the farmers have grown bananas, cassava, pawpaws, sugar cane and fodder. These crops have performed well throughout the dry season which demonstrates that the seasonal residual moisture can be used productively.

High evaporation losses have been estimated in the study. The sharp decline in moisture levels, particularly during dry seasons when the soil is bare, imply that this loss of moisture to unproductive purposes reduce the initial moisture conditions on the onset of the following season. This compromises the length of the growing seasons as more water would be required to bring soil moisture levels to acceptable levels for crop growth. Reducing this direct evaporation through options such as mulching would help to minimise soil moisture loss.

During the research complete crop failure has been observed in a very dry season (*Masika* 2007) while all sites performed extremely well in a very wet season (*Vuli* 2006). In average seasons, the sites with better diversions and conservation agriculture performed better than others. It can therefore be concluded that the tested techniques will not provide total guarantee of yields in extremely dry seasons. The impact of the techniques is not noticeable either when the rainfall distribution is good. However, in average and slightly below average seasons, which are common in SSA, the tested techniques are effective.

This research shows that there is scope to improve the current yield levels provided appropriate alternative practices are adopted. In this research, emphasis has been placed on techniques which can be afforded by the majority of the vulnerable rural communities. If, for instance, the farmers invest in water security and nutrient management, the yield levels recorded here would certainly further improve.

The next important step is to see the communities shifting to more efficient technologies after witnessing such successes. However, once-off demonstrations are not sufficient and more similar research, spread over wider areas, would help to disseminate the findings and to reach more solid conclusions after wider research. Research would need to be coupled with parallel efforts in policy shifts and the creation of markets to attract communities to adopt more efficient practices. For instance, issues of adoption cannot be measured accurately within the period of the research. It would be expected that adoption and adaption takes place after the lifetime of the research. On the ground, extention services which are up to date with scientific advancement would help to guide farmers. Similarly, while it is believed that the *fanya juus*s result in the deposition of fine, nutrient rich sediments, it is also not possible to measure these changes within the 4 years of this research.

This research has explained the repartitioning of water using different scientific approaches (direct observations, geophysical techniques and modelling). It has not been possible in this research to upscale these local successes to catchment scales under similar environments. This raises the question of how far such successes can be sustained within a catchment without generating conflicts with other competing uses. Follow up multi-disciplinary research is therefore recommended to identify the longer term impacts at local scales and to downstream ecosystems.

REFERENCES

Agarwal, A., Narain, S. and Khurana, I. (Editors), 2005. Making Water Everybody's Business. Centre for Science and Environment.

Ali, M.H. and Talukder, M.S.U., 2008. Increasing water productivity in crop production - a synthesis. Agricultural Water Management, 95: 1201-1213.

Allen, R., Pereira, L., Smith, M., Raes, D. and Wright, J., 1998. Crop Evapotranspiration: Guidelines for computing crop water requirements. FAO Irrigation and Drainage Paper, 56.

Allen, R., Pereira, L., Smith, M., Raes, D. and Wright, J., 2005. FAO-56 Dual Crop Coefficient Method for Estimating Evaporation from Soil and Application Extensions. Journal of Irrigation and Drainage Engineering.

Allen, R.G., 2000. Using the FAO-56 dual crop coefficient method over an irrigated region as part of evapotranspiration intercomparison study. Journal of Hydrology 22, 229: 27-41.

Asfahani, J., 2007. Geoelectrical investigation for characterizing the hydrogeological conditions in semi-arid region in Khanasser valley, Syria. Journal of Arid Environments 68: 31–52.

Barron, J. and Rockström, J., 2003. Water harvesting to upgrade smallholder farming. Experiences from on-farm research in Kenya and Burkina Faso. RELMA Publication.

Barron, J., Rockström, J., Gichuki, F. and Hatibu, N., 2003. Dry spell analysis and maize yields for two semi-arid locations in East Africa. Agricultural and Forest Meteorology, 117: 23-37.

Bewket, W. and Sterk, G., 2002. Farmers' participation in soil and water conservation activities in the Chemoga Watershed, Blue Nile Basin, Ethiopia. Land Degradation and Development, 13: 189-200.

Bhatt, Y., Bossio, D., Enfors, E., Gordon, L., Kongo, V., Kosgei, J.R., Makurira, H., Masuki. K., Mul, M., and Tumbo, S.D., 2006. Smallholder system innovations in integrated watershed management (SSI): Strategies of water for food and environmental security in drought-prone tropical and subtropical agro-ecosystems, IWMI Colombo, Sri Lanka.Binley, A., Slater, L.D., Fukes, M. and Cassiani, G., 2005. Relationship between spectral induced polarization and hydraulic properties of saturated and unsaturated sandstone. Water Resources Research, 41(W12417, doi:10.1029/2005WR004202).

Chikozho, C., 2005. Policy and institutional dimensions of small-holder farmer innovations in the Thukela River Basin of South Africa and the Pangani River Basin of Tanzania. A comparative perspective. Physics and Chemistry of the Earth, 30: 913-924.

Critchley, W. and Siegert, K., 1991. Water Harvesting: A manual for the design and construction of water harvesting schemes for plant production, FAO, Rome.

Dabrowski, J.M., Masekoameng, E. and Ashton, P.J., 2009. Analysis of virtual water flows associated with the trade of maize in the SADC region. Hydrol. Earth Syst. Sci., 13: 1967-1977.

de Fraiture, C., Wichelns, D., Rockström, J. and Kemp-Benedict, E. (Editors), 2007. Looking ahead to 2050. Scenarios of alternative investment approaches, Chapter 3. IWMI, Colombo.

De Groen, M.M. and Savenije, H.H.G., 2006. A monthly interception equation based on the statistical characteristics of daily rainfall Water Resources Research, 42(W12417): 1-10.

Decagon Devices Inc., 2004. AccuPar PAR/LAI ceptometer, Model LP-80: Operators Manual. Version 1.2.

Dlamini P., Orchard, C., Jewitt, G., Lorentz, S., Titshall, L., and Chaplot, V., submitted. Quantifying interril erosion in a sloping-land-agricultural catchment of KwaZulu Natal, South Africa. Agricultural Water Management.

Enfors, E., 2009. Traps and transformations: Exploring the potential of water system innovations in dryland sub-Saharan Africa. PhD Thesis, Stockholm University, Stockholm.

Enfors, E., Barron, J., Makurira, H., Rockström, J. and Tumbo, S.D., 2010. Yield and soil system changes from conservation tillage in dryland farming: a case study from North Eastern Tanzania. Agricultural Water Management, doi:10.1016/j.agwat.2010.02.013.

Enfors, E. and Gordon, L., 2007. Analyzing resilience in dryland agro-ecosystems. A case study of the Makanya catchment in Tanzania over the past 50 years. Land degradation and development, 18: 680-696.

Enfors, E. and Gordon, L., 2008. Dealing with drought: The challenge of using water system technologies to break dryland poverty traps. Global Environmental Change, 18(4): 607-616.

Enfors, E., Gordon, L., Peterson, G.D. and Bossio, D., 2008. Making dryland development work: participatory scenario planning to increase the robustness of investments in small-scale farming in the Makanya catchment, Tanzania. Ecology and Society, 13(2): 42.

Falkenmark, M., 1995. Coping with Water Scarcity under Rapid Population Growth, Conference of SADC Ministers, Pretoria.

Falkenmark, M. and Molden, D., 2008. Wake up to realities of river basin closures. Water Resources Development, 24(2): 201-215.

FAO, 2001. The economics of conservation agriculture. FAO, Rome.

FAO, 2002. The state of food insecurity in the world, Rome.

FAO, 2003. Water for sustainable food production, poverty alleviation and rural development, Rome.

FAO, 2005. FAOSTAT (Food and Agriculture Organisation Statistics).

FAO, 2008. State of food insecurity in the world, Rome.

Fischer, B.M.C., 2008. Spatial variability of dry spells, a spatial and temporal rainfall analysis of the Pangani Basin & Makanya catchment, Tanzania. MSc Thesis, Delft University of Technology, Delft.

Gichuki, F., 2000. Mokueni District Profile: Soil Management and Conservation. Drylands Research, Working Paper 4.

Giller, K.E., Witter, E., Corbeels, M. and Tittonell, P., 2009. Conservation agriculture and smallholder farming in Africa: The heretics' view. Field Crops Research 114, 114: 23-34.

Hatibu, N., 2002. Rainwater Management: Strategies for Improving Water Availability and Productivity in Semi-arid and Arid Areas, Issues of water management in agriculture: a compilation of essays. IWMI.

Hatibu, N. and Mahoo, H.F., 1999. Rainwater harvesting technologies for agricultural production: A case for Dodoma, Tanzania. Conservation tillage with animal traction, A resource book of ATNESA.

Hatibu, N., Mutabazi, K., Senkondo, E.M. and Msangi, A.S.K., 2006. Economics of rainwater harvesting for crop enterprises in semi arid areas of East Africa. Agricultural Water Management, 80(74-86).

Hoff, H., Falkenmark, M., Gerten, D., Gordon, L., Karlberg, L., and Rockström, J., 2009. Greening the global water system. Journal of Hydrology, In Press (doi:10:1016/j.hydrol.2009.06.026).

Humphreys, E., Peden, D., Twomlow, S., Rockström, J., Oweis, T., Huber-Lee, A., and Harrington, L., 2008. Improving rainwater productivity: Topic 1 Synthesis paper. CGIAR Challenge Programme on Water and Food, Colombo: 19pp.

Igbadun, H.E., Mahoo, H.F., Tarimo, A.K.P.R. and Salim, B.A., 2006. Crop water productivity of an irrigated maize crop in Mkoji sub-catchment of the Great Ruaha River Basin, Tanzania. Agricultural Water Management, 85: 141-150.

IWMI, 2003. Innovative Approaches to Agricultural Water Use for Improving Food Security in Sub-Saharan Africa.

Jewitt, G., 2006. Integrating blue and green water flows for water resources management and planning. Physics and Chemistry of the Earth, 31: 753-762.

Kemerink, J.S., Ahlers, R. and Van der Zaag, P., 2009. Assessment of the potential of hydro-solidarity within plural condition of traditional irrigation systems in northern Tanzania. Physics and Chemistry of the Earth, 34: 881-889.

Kijne, J.W., Barron. J., Hoff, H., Rockström, J., Karlberg, L., Gowing, J., Wani, S. P., and Wilchens, D., 2009. Opportunities to increase water productivity in agriculture with special reference to Africa and South Asia, Stockholm Environmental Institute, Stockholm.

Kimaro, A.A., Timmer, V.R., Chamshama, S.A.O., Ngaga, Y.N. and Kimaro, D.A., 2009. Competition between maize and pigeonpea in semi-arid Tanzania: Effect on yields and nutrition of crops. Agriculture, Ecosystems and Environment, 134(115-125).

Koch, K., Wenninger, J., Uhlenbrook, S. and Bonnell, M., 2008. Electrical resistivity tomography (ERT) for identifying hillslope processes in the Black Forest Mountains, Germany. Hydrological Processes.

Komakech, C.H., Mul, M.L., van der Zaag, P. and Rwehumbiza, F.B.R., submitted. Potential of spate irrigation in improving rural livelihoods: case of Makanya spate irrigation system, Tanzania. Agricultural Water Management.

Kongo, M.V., 2008. Balancing Water for Food and Environment: Hydrological Determinants across scales in the Thukela river basin. PhD Thesis, University of KwaZulu_Natal, South Africa.

Kosgei, J.R., 2009. Rainwater harvesting systems and their influences on field scale soil hydraulic properties, water fluxes and crop production in Potshini catchment, South Africa. PhD Thesis, University of KwaZulu-Natal, South Africa.

Kosgei, J.R., Jewitt, G.P.W., Kongo, V.M. and Lorentz, S.A., 2007. The influence of tillage on field scale water fluxes and maize yields in semi-arid environments: A case study of Potshini catchment, South Africa. Physics and Chemistry of the Earth, 32(15-18): 1117-1126.

Liu, J., Zehnder, A.J.B. and Yang, H., 2008. Drops for crops: Modelling crop water productivity on a global scale. Global Nest Journal, 10(3): 295-300.

Loke, H.M., 2003. Tutorial: 2-D and 3-D Electrical Imaging Surveys.

Makurira, H., Mul, M.L., Vyagusa, N.F., Uhlenbrook, S. and Savenije, H.H.G., 2007a. Evaluation of community-driven smallholder irrigation in dryland South Pare Mountains, Tanzania: A case study of Manoo micro dam. Physics and Chemistry of the Earth 32(15-18): 1090-1097.

Makurira, H., Savenije, H., Uhlenbrook, S., Rockström, J. and Senzanje, A., submitted. The effect of improved on-farm techniques on maize yield and water productivity in sub-Saharan rainfed agricultural systems. Journal of Agricultural Water Management.

Makurira, H., Savenije, H.H.G. and Uhlenbrook, S., 2007b. Towards a better understanding of water partitioning processes for improved smallholder rainfed agricultural systems: A case study of Makanya catchment, Tanzania. Physics and Chemistry of the Earth, 32(15-18): 1082-1089.

Makurira, H., Savenije, H.H.G. and Uhlenbrook, S., 2009a. Modelling field scale water partitioning using on-site observations in sub-Saharan rainfed agriculture. Hydrol. Earth Syst. Sci. Discuss., 6: 5537-5563.

Makurira, H., Savenije, H.H.G., Uhlenbrook, S., Rockström, J. and Senzanje, A., 2009b. Investigating the water balance of on-farm techniques for improved crop productivity in rainfed systems: A case study of Makanya catchment, Tanzania. Physics and Chemistry of the Earth, 34: 93-98.

Mazvimavi, D., Madamombe, E.K. and Makurira, H., 2007. Assessment of environmental flow requirements for river basin planning in Zimbabwe. Physics and Chemistry of the Earth, 32: 995-1006.

Mbugua, J. and Nissen-Petersen, E. (Editors), 1995. Rainwater: An underutilised resource. SIDA.

Mirghani, M.M.O. and Savenije, H.H.G., 1995. Incorporation of People's Participation in Planning and Implementation of Water Resources Projects. Physics and Chemistry of the Earth, 20(3-4): 229-236.

Motsi, K., Chuma, E. and Mukamuri, B.B., 2004. Rainwater harvesting for sustainable agriculture in communal lands of Zimbabwe. Physics and Chemistry of the Earth, 29: 1069-1073.

Mugabe, F.T., Senzanje, A. and Hodnett, M., 2002. Understanding catchment hydrology as a means to improve rural water resources and management in Masvingo Province., Workshop on the processes and dynamics of catchment management in Zimbabwe. Save African Trust.

Mul, M.L., 2009. Understanding hydrological processes in an ungauged catchment in sub-Saharan Africa. PhD Thesis, UNESCO-IHE, Delft, 130 pp.

Mul, M.L., Kemerink, J.S., van der Zaag, P., Vyagusa, F.N., Mshana, M.G., and Makurira, H., 2010. Water allocation practices among smallholder farmers in South Pare Mountains, Tanzania; The issue of scales. Journal of Agricultural Water Management, doi:10.1016/j.agwat.2010.02.014

Mul, M.L., Mutiibwa, R.K., Foppen, J.W.A., Uhlenbrook, S. and Savenije, H.H.G., 2007. Identification of groundwater flow systems using geological mapping and chemical spring analysis in South Pare Mountains, Tanzania. Physics and Chemistry of the Earth, 32(15-18): 1015-1022.

Mul, M.L., Savenije, H.H.G. and Uhlenbrook, S., 2008. Understanding key hydrological processes in a small semi-arid catchment in Tanzania by using a multi-method approach, WaterNet/WARFSA/GWP-SA annual symposium, Johannesburg, South Africa.

Munodawafa, A. and Zhou, N., 2008. Improving water utilization in maize production through conservation tillage systems in semi-arid Zimbabwe. . Physics and Chemistry of the Earth, 33: 757-761.

Mupangwa, W., Love, D. and Twomlow, S., 2006. Soil water conservation and rainwater harvesting strategies in the semi arid Mzingwane Catchment, Limpopo Basin, Zimbabwe. Physics and Chemistry of the Earth, 31: 893-900.

Mutiro, J., Makurira, H., Senzanje, A. and Mul, M.L., 2006. Water productivity analysis for smallholder rainfed systems: A case study of Makanya catchment, Tanzania. Physics and Chemistry of the Earth, 31(15-16): 901-909.

Mwamfupe, D.G., 2002. The role of non-farm activities in household economies: a case study of Pangani Basin, Tanzania. In: J.O. Ngana (Editor), Water resources management: the case of Pangani River Basin. Issues and approaches. Dar es Salaam University Press, Dar es Salaam, pp. 39 – 47.

Mwangi, J.N., Mboya, T.O. and Kihumba, J., 2001. Improved maize production in Central Kenya with adoption of soil and water conservation measures, Proceedings of the 7th Eastern and Southern African Regional Maize Conference, pp. 299-300.

Mwenge Kahinda, J.-M., Rockström, J., Taigbenu, A.E. and Dimes, J., 2007. Rainwater harvesting to enhance water productivity of rainfed agriculture in the semi-arid Zimbabwe. Physics and Chemistry of the Earth, 32(15-18): 1068-1073.

Ndiaye, B., Mole´nat a, J., Hallaire, V., Gascuel, C. and Hamon, Y., 2007. Effects of agricultural practices on hydraulic properties and water movement in soils in Brittany (France). Soil & Tillage Research, 93: 251-263.

Ngigi, S.N. (Editor), 2003. Rainwater Harvesting for Improved Food Security. Promising Technologies in the Great Horn of Africa. Greater Horn of Africa Partnerships. Kenya Rainwater Association.

Ngigi, S.N., Savenije, H.H.G. and Gichuki, F.N., 2007. Land use changes and hydrological impacts related to up-scaling of rainwater harvesting and management in upper Ewaso Ng'iro river basin, Kenya. Land Use Policy, 24: 129-140.

Novák, V., Hurtalová, T. and Matejka, F., 2005. Predicting the effects of soil water content and soil water potential on transpiration of maize. Agricultural Water Management, 76: 211-223.

Nyagumbo, I., Munamati, M., Chikwari, E. and Gumbo, D., 2009. In-situ water harvesting technologies in semi-arid southern Zimbabwe: Part I. The role of biophysical factors on performance in Gwanda District, 10th Waternet Warfsa Symposium, Entebbe, Uganda.

Pacey, A. and Cullis, A. (Editors), 1986. Rainwater Harvesting: The collection of rainfall and runoff in rural areas. Intermediate Technology Publications.

Pachepsky, Y. Guber, A., Jacques, D., Simunek, J., van Genuchten, M., Nicholson, T., and Cady, R., 2006. Information content and complexity of simulated soil water fluxes. Georderma, 134: 253-266.

Pangani Basin Water Office, 2005. Unpublished office data.

Reij, C., Mulder, P. and Begemann, L., 2001. Water balance accounting for designing and planning rainwater-harvesting systems for supplementary irrigation.

Rockström, Hatibu, N., Oweis, T. and Wani, S.P., 2007. Managing water in rainfed agriculture: In water for food, water for life. A comprehensive Assessment of Water Management in Agriculture (Earthscan, London and IWMI Colombo, Sri Lanka): 315-348.

Rockström, J., 2000. Water resources management in smallholder farms in eastern and southern Africa: An overview. Physics and Chemistry of the Earth, 25(3): 275-283.

Rockström, J., 2001. Green water security for the food makers of tomorrow: windows of opportunity in drought-prone savannahs. Water Science and Technology, 43(4): 71-78.

Rockström, J., 2003a. Resilience building and water demand management for drought mitigation. Physics and Chemistry of the Earth, 28: 869-877.

Rockström, J., 2003b. Water for food and nature in drought-prone tropics: vapour shift in rainfed agriculture. The Royal Society: 1997 - 2009.

Rockström, J., Folke, C., Gordon, L., Hatibu, N., Jewitt, G., Rwehumbiza, F., Sally, H., Savenije, H., and Schulze, R., 2004. A watershed approach to upgrade rainfed agriculture in water scarce regions through water systems innovations: an integrated research initiative on water for food and rural livelihoods in balance with ecosystem functions. Physics and Chemistry of the Earth, 29: 1109-1118.

Rockström, J., Falkenmark, M., Karlberg, L., Hoff, H., Rost, S., and Gerten, D., 2009. Future water availability for global food production: The potentialof green water for increasing resilience to global change, Water Resources Research.

Rockström, J., Kaumbutho, P., Mwalley, P. and Temesgen, M., 2001. Conservation farming among smallholder farmers in E.Africa: Adapting and adopting innovative land management options. 1st World Congress on Conservation Agriculture: 363-373.

Savenije, H.H.G., 1997. Determination of evaporation from a catchment water balance at a monthly time scale. hydrological Earth System Sciences, 1: 93-100.

Savenije, H.H.G., 1998. "How do we feed a growing world population in a situation of water scarcity?"; in "Water-The Key to Socio-economic Development and Quality of Life", 8th Stockholm Symposium, 10-13 August, 1998. SIWI, pp. 49-58.

Savenije, H.H.G., 2000. Water scarcity indicators; the deception of the numbers. Physics and Chemistry of the Earth, 25(3): 199-204.

Simunek, J., van Genuchten, M.T. and Sejna, M., 2006. The HYDRUS software package for simulating two and three dimensional movement of water, heat, and multiple solutes in variably saturated media, version 1.0, Prague, Czech Republic.

SSI, 2002. Project proposal.

Swatuk, L.A. and Motsholapheko, M., 2008. Communicating integrated water resources management: From global discourse to local practice - Chronicling an experience from Boteti River sub-basin, Botswana. Physics and Chemistry of the Earth, 33: 881-888.

Temesgen, M., Rockström, J., Savenije, H.H.G. and Hoogmoed, W.B., 2007. Assessment of strip tillage systems for maize production in semi-arid Ethiopia: Effects on grain yield and water balance. HESSD, 4: 2229-2271.

Tenge, A.J.M., De Graaff, J. and Hella, J.P., 2005. Financial efficiency of major soil and water conservation measures in West Usambara highlands, Tanzania. Applied Geography, 25: 348-366.

Tumbo, S.D., Mutabazi, K.D., Byakugila, M.M. and Mahoo, H.F., 2010. An empirical framework for scaling-out of water system innovations: lessons from diffusion of terraces in the Makanya catchment in northern Tanzania. Agric. Water Manage., doi:10.1016/j.agwat.2010.03.010.

Uhlenbrook, S., 2007. Biofuel and water cycle dynamics: what are the related changes for the hydrological processes research? Hydrological Processes, 21: 3647-3650.

UN, 2005. The Millennium Development Goals Report 2005. United Nations, Department of Public Information.

UN, 2008. The Millennium Goals Development Report.

UNFPA, 2008. Global regional programme 2008-2011, United Nations Population Fund.

URT, 2004. 2002 Population and Housing Census, Volume IV, District profile, Same, Central Census Office, National Bureau of Statistics, President's Office, Planning and Privatisation, Dar Es Salaam.

URT, 2005. Tanzania Poverty and Human Development Report. Ministry of Planning, Economy and Empowerment, Dar es Salaam.

van der Giesen, N.C., Stomph, T.J. and de Ridder, N., 2000. Scale effects of Hortonian overland flow and rainfall-runoff dynamics in a West African catena landscape. Hydrological Processes, 14(1): 165-175.

Van der Zaag, P., 2005. Integrated water resources management: relevant concept or irrelevant buzzword? A capacity building and research agenda for southern Africa. Physics and Chemistry of the Earth 30: 867-871.

Verbist, K., Cornelis, W.M., Gabriels, D., Alaerts, K. and Soto, G., 2009. Using an inverse modelling approach to evaluate the water retention in a simple water harvesting technique. Hydrol. Earth Syst. Sci., 13: 1979-1992.

Vishnudas, S., 2006. Sustainable water management: An illusion or reality? A case of Kerala State in India. PhD Thesis, Delft University of Technology, The Netherlands, Delft.

Wallace, J.S., Jackson, N.A. and Ong, C.K., 1999. Modelling soil evaporation in an agroforestry system in Kenya. Agricultural and Forestry Meteorology, 94: 189-202.

Wenninger, J., Uhlenbrook, S., Lorentz, S. and Leibundgut, C., 2008. Identification of runoff generating processes using combined hydrometric, tracer and geophysical methods in a headwater catchment in South Africa. Hydrological Sciences Journal, 53(1): 65-80.

Zwart, S.J. and Bastiaanssen, W.M.G., 2004. Review of measured crop water productivity values for irrigated wheat, rice, cotton and maize. Agricultural Water Management, 69: 115-133.

SAMENVATTING

Het produceren van voldoende voedsel is een groot probleem in sub-Sahara Afrika (SSA) waar, veelal door zeer variabele regenval in combinatie met hoge verdamping en afnemende bodemkwaliteit, de oogstopbrengsten achteruit gaan. Het vocht- en nutriëntgehalte in de grond zijn essentieel voor een goede oogst. Watertekort wordt gezien als de meest beperkende factor voor gewasproduktie (CP) in SSA.

Het probleem om honger en armoede in SSA te bestrijden zal zeer groot blijven door de bevolkinggroei van 2.2% a^{-1} en een verwachte bevolkingsomvang van 1.1 biljoen in 2010, tenzij nieuwe ingrepen het tij kunnen keren. De oogstopbrengsten in SSA zijn laag, een gemiddelde maïsoogst komt nauwelijks boven de 1 t ha^{-1}, wat erg laag is vergeleken met de potentiële oogst van 6 t ha^{-1}.

Het watertekort voor de landbouw is in het verleden altijd opgelost door het ontwikkelen van nieuwe irrigatiegebieden. Echter, de irrigatietechniek is niet aan iedereen ten goede gekomen, door de hoge investeringskosten die nodig zijn om een irrigatiesysteem te bouwen. Hierdoor is het overgrote deel van de bevolking in SSA (80-90%) nog steeds zuiver afhankelijk van regenval voor hun levensonderhoud.

Regenval in SSA is in het algemeen onbetrouwbaar, met seizoensregenval variërend tussen de 300 en 1200 mm a^{-1}. Ook binnen één seizoen varieert de regenval sterk, evenals de potentiële verdamping, die in vele locaties de 1000 mm a^{-1} overschrijdt. Dit betekent dat elke regendruppel die valt productief gebruikt zal moeten worden om de gewasopbrengsten te verhogen. Het probleem van het watertekort dat ontstaat door onvoldoende seizoensregenval en opeenvolgende periodes van droogte wordt verder versterkt door de inefficiënte landbouwpraktijken van de kleinschalige boeren, die onvoldoende bodembeschermende maatregelen toepassen. Wanneer de periodes van droogte optreden tijdens de kritieke groeimomenten van de gewassen zal dit de oogst sterk reduceren, zelfs als de gemiddelde regenval dat seizoen als voldoende beschouwd kan worden.

De boeren hebben ook te weinig middelen om de juiste producten te kopen die ervoor kunnen zorgen dat het gewas het water- en nutriëntentekort kan overbruggen. Opmerkelijk is dat veel onderzoek is gedaan naar nieuwe methodes om beter om te gaan met periodes van droogte, evenals naar bodembeschermende en waterbesparende technieken en het beheer van de akkers. Echter, in de praktijk is de situatie niet verbeterd, en de oogstopbrengsten nemen zelfs af. Tot op heden blijkt er een probleem te zijn om, met behulp van onderzoek, tot duurzame oplossingen te komen voor de blijvend lage productiviteit van de kleinschalige, regenvalafhankelijke boeren. Dit onderzoek hoopt een bijdrage te leveren aan het oplossen van dit probleem.

De hypothese van dit onderzoek is dat veel van het onderzoek in het verleden te gefragmenteerd was om de beschreven problemen op te lossen. Het is moeilijk om zulke gefragmenteerde aanpakken in de praktijk toe te passen. Een holistische aanpak om traditionele landbouwsystemen te ondersteunen zal zowel de hydrologische, agronomische, bodemtechnische, stroomgebied-beschermende en sociaal-economische factoren moeten bestuderen om succesvol te zijn.

Dit onderzoek is gedaan in het Makanya stroomgebied in het noorden van Tanzania. De gemiddelde regenval in het gebied is minder dan 400 mm seizoen^{-1}, wat beduidend minder is dan de benodigde hoeveelheid water voor de voorkeursgewassen zoals maïs. Innovaties (SIs) voor alternatieve landbouwpraktijken zijn hier geïntroduceerd en getest. Deze innovaties behelzen een combinatie van duurzame landbouw, het aftappen van oppervlakteafvoer naar de akkers en verbetering van het lokaal bodemvochtgehalte door middel van het graven van greppels met bovenstroomse wallen (*fanya juus*) binnen de bewerkte akkers. Deze technieken zijn geselecteerd voor het onderzoek omdat ze aan de ene kant geen hoge investeringskosten vereisen en daarom betaalbaar zijn voor veel boeren en aan de andere kant de oogstopbrengsten significant kunnen verhogen.

Uitgebreid participatief veldonderzoek is gedaan naar de regenval, bodemverdamping, afvoerbijdrage, seizoenopbrengst van maïs en het algemene functioneren van de gewassen. Indirecte methodes zijn toegepast om de veldobservaties te bevestigen en tevens om de prestaties van het bestudeerde systeem te modeleren. Elektrische Weerstand Tomografie (ERT) is gebruikt als geofysische techniek om het bodemvochtgehalte bepaald met de Tijds-Domein-Reflectometer (TDR) te bevestigen. Eveneens is het HYDRUS2D model toegepast om de tweedimensionale laterale grondwaterstroming te simuleren, aan de hand van het water dat op de meetlocaties is toegevoegd.

Om beter inzicht te krijgen in de waterverdeling onder verschillende scenario's en om het effect op de gewassen te kunnen kwantificeren is een spreadsheet model gemaakt gebaseerd op de waterbalans. Resultaten van het onderzoek laten zien dat regenval gemiddeld laag is en varieert tussen de 150 - 300 mm/seizoen gedurende de onderzoeksperiode. De variabiliteit van de regenval is groot tussen de seizoenen en tussen de verschillende meetlocaties. De opbrengsten van de maïsoogsten zijn ook laag en variëren van een gemiddeld minimum van 0.23 t ha^{-1} (gebruikmakend van de huidige landbouwtechnieken) tot een maximum van 2.82 t ha^{-1} (bij toepassing van de SIs op alle meetlocaties). Het beste resultaat van 4.8 t ha^{-1} werd behaald in een goed seizoen op een meetlocatie waar, naast de regenval, een combinatie werd toegepast van het aftappen van oppervlakte afvoeren naar de akkers, het land bewerken met een diepe ploeg, het verbeteren van het bodemvochtgehalte gebruikmakend van *fanya juus* en het aanbrengen van mest.

Analyse van de waterverdeling bevestigt dat de toegepaste SIs tot een toename van de transpiratie leidt van ongeveer 49%. Deze toename in efficiëntie kan worden uitgelegd door zowel directe als indirecte analytische technieken toe te passen. Om beter inzicht te krijgen in de interacties tussen het water en de bodem zijn binnen het onderzoek verschillende analytische technieken toegepast op veldschaal. Het onderzoek toont aan dat er inderdaad ruimte is om de productiviteit van het water te vergroten indien de lokale boeren efficiëntere landbouwbewerking technieken toepassen.

Door meer water af te tappen neemt de oogstopbrengst significant toe, en dit effect wordt verder versterkt indien ook andere SIs technieken worden toegepast zoals diep ploegen, bemesting, en het afdekken van de grond rond de gewassen. Dit bevestigt dat er meerdere maatregelen nodig zijn om het probleem van de lage oogstopbrengst van kleinschalige boeren op te lossen. De geteste SIs dragen bij aan een grotere voedselzekerheid door een toename in de oogstopbrengsten. Bovendien bieden ze de mogelijkheid om verschillende gewassen tegelijkertijd te verbouwen doordat de landbouwtechnieken een heterogene situatie creëren in de akker door de verandering in de waterbalans.

Het onderzoek toont echter ook aan dat zelfs met deze veel belovende resultaten er nog ruimte is voor verdere verbetering van de waterefficiëntie voor het groeien van gewassen door verbeterde onderzoeksmethodes toe te passen en betere technieken te ontwikkelen.

About the Author

Hodson Makurira was born in 1966 in Bindura, Zimbabwe. He obtained his Bachelor of Science degree in Civil Engineering from the University of Zimbabwe. Between 1992 and 2002 he worked as a water resources engineer for the Ministry of Water Resources and Development and, later, the Zimbabwe National Water Authority (ZINWA). During the period 1995-1997 he studied in Delft at the International Institute for Infrastructural, Hydraulic and Environmental Engineering (now UNESCO-IHE) for his Master of Science Degree in Water and Environmental Resources Management (WREM).

In 2003, Hodson joined the Civil Engineering Department at the University of Zimbabwe as a lecturer in the Water Section. In 2004, he joined the Smallholder Systems Innovations (SSI) Programme as a PhD researcher working in a multi-disciplinary research team aimed at improving livelihoods for communities living in semi-arid rainfed environments. His part of the research within the multi-disciplinary SSI programme focused on proposing and testing innovative techniques to mitigate the effects of dry spells on crop productivity. His research sought to better understand water partitioning processes at field scale and linking these processes to crop productivity.

While pursuing his PhD studies, Hodson was also actively involved in the delivery of the WaterNet-sponsored Masters Programme in Integrated Water Resources Management (IWRM). He supervised several Masters theses and was involved in diverse research activities leading to nine publications in peer-reviewed journals.

Hodson completed his PhD studies in June 2010.

For Product Safety Concerns and Information please contact our EU representative GPSR@taylorandfrancis.com Taylor & Francis Verlag GmbH, Kaufingerstraße 24, 80331 München, Germany

Batch number: 10722120

Printed by Printforce, the Netherlands